(Limited to One Thousand Copies.)

HISTORICAL RECORD

AND

REGIMENTAL MEMOIR

OF

THE ROYAL SCOTS FUSILIERS

FORMERLY KNOWN AS THE

21st Royal North British Fusiliers.

Containing an Account of the Formation of the Regiment in 1678 and its subsequent Services until June 1885.

COMPILED FROM VARIOUS AUTHENTIC SOURCES

BY

JAMES CLARK,

LATE SERGEANT ROYAL SCOTS FUSILIERS.

ILLUSTRATED WITH SIX COLOURED PLATES.

THE ROYAL SCOTS FUSILIERS

BEARS ON

The Regimental Colour

THE THISTLE

Within the circle, and Motto of Saint Andrew,

"*Nemo me impune lacessit*,"

Surmounted by the Imperial Crown with the Royal Cipher, and a Crown in each of its three corners.

The Regimental Colour

ALSO BEARS THE FOLLOWING HONOURS,

"*Blenheim.*"	"*Ramillies.*"	"*Oudenarde.*"
"*Malplaquet.*"	"*Deddingen.*"	"*Bladensburg.*"
"*Alma.*"	"*Inkerman.*"	"*Sevastopol.*"

"*South Africa, 1879.*"

Regimental Badges.

A BURSTING GRENADE, WITH ROYAL ARMS ON THE BALL.
That on the Collar bears a Thistle.

Uniform.

SCARLET HIGHLAND DOUBLET, FACINGS BLUE, TARTAN TREWS, AND SEALSKIN BUSBY.

By Royal Warrant of 1st July 1751, *the Thistle as on the Colours, the White Horse and motto over it*, "*Nec aspera terrent*," were authorised to be worn on the grenadier caps; the device of *Thistle and Crown* to be painted on bells of arms and drums.

TO

THE OFFICERS, NON-COMMISSIONED OFFICERS, AND SOLDIERS

OF

THE ROYAL SCOTS FUSILIERS

PAST, PRESENT, AND FUTURE,

THE FOLLOWING RECORD OF THE SERVICES OF THEIR DISTINGUISHED OLD CORPS

IS

Most Respectfully Dedicated

BY THEIR OBEDIENT SERVANT,

THE COMPILER.

Preface.

THERE exists in the breasts of most of those who have served, or are serving, in the army, an *esprit de corps*,—an attachment to everything belonging to their regiment; to such, a narrative of the services of their own corps cannot fail to prove interesting. Authentic accounts of the actions of the great, the valiant, the loyal, have always been of paramount interest with a brave and civilised people. Great Britain has produced a race of heroes, who, in moments of danger and terror, have stood "firm as the rocks of their native shore;" and, when half the world has been arrayed against them, they have fought the battles of their country with unshaken fortitude. It is presumed that a record of achievements in war,—victories so complete and surprising, gained by our countrymen, our brothers, our fellow-citizens in arms,—a record which revives the memory of the brave, and brings their gallant deeds before us, will certainly prove acceptable to the public.

From the materials thus collected in the following pages readers will be able to realise the difficulties and privations which chequer the career of those who embrace the military profession. In Great Britain comparatively little is known of the vicissitudes of active service, and of the casualties of climate, to which, even in peace, the British troops are exposed in every part of the globe, with little or no interval of repose.*

Although a long time has elapsed since my connection with the Royal Scots Fusiliers ceased, my recollections of the regiment are fresh and pleasant, and my affection for it remains undiminished. Nor am I singular in my love for the "old corps;" all who have had the good fortune to belong to it, I believe, are animated with the same feeling.

The cause of this, doubtless, is to be found in the admirable system which prevailed in the regiment,—a system necessarily strict, but equitable, and always kindly administered.

Every man felt that he belonged to a *large family*, in which the personal rights and interest of each member were respected and held sacred.

The successive commanding-officers, especially those whose military services were confined entirely to the regiment, took a personal concern in everything connected with the well-being of all under their command. It cannot therefore be wondered at, that those who have served in *such a regiment* should continue to entertain for it feelings of deep affection.

The greater portion of the following pages have, thanks to the kindness of the publishers and editor, already appeared in the columns of the *Ayrshire Post*, to which has now been added a short but comprehensive sketch of the services of the second battalion of the Fusiliers, with

* From Cannon's *Military Records*.

other additional interesting particulars embodied in "Appendices."

I am under much obligation to A. Ross, Esq., S.S.C., Edinburgh, who has kindly permitted me to make use of any item referring to the Royal Scots Fusiliers in the pages of his handsome addition to our military literature, *The Old Scottish Regimental Colours*;* Cannon's *Historical Record* up to 1849, and Kinglake's *Invasion of the Crimea*,† have been also quoted; also details from subsequent official sources, newspaper cuttings, and letters, personal reminiscences, verbal and written contributions from "Old Fusiliers,"—all have been incorporated to form, I trust, a symmetrical whole.

I am also under the deepest obligation to an old Fusilier, Colour-Sergeant W. Heaney, of the first battalion, whose literary acquirements, long connection with and intimate knowledge of the regiment, eminently fitted him for the task he has so generously fulfilled in editing this volume. Without his valuable and most friendly aid, I should not have been able to present this book to the public in so complete a form.

I beg to assure those who have kindly furnished matter for insertion in these records, and who on reading the book

---

* The Royal Scots Fusiliers are referred to in the following pages of *The Old Scottish Regimental Colours*:—5 note, 10, 11, 13, 16–20, 21 note, 22, 24 note, 28 note, 32 note, 34 note, 39, 41 note, 60–63, 69, 71, 72, 78 note, 83, 134, 135.

† Kinglake, in his *Invasion of the Crimea*, refers particularly to the Royal Scots Fusiliers in the following pages of his exhaustive and authoritative work:—Vol. iii., p. 95; vol. v., chap. vi., pp. 210, 235–237, 366, 367, 369, 370, 419, 422–424.

find that their contributions have been curtailed, condensed, or wholly omitted, that these curtailments, condensations, and omissions have not been made arbitrarily nor capriciously, but from the belief that they were necessary.

To put before the reader *facts*, concisely and clearly stated, is what has been sought to be accomplished by the Editor.

In conclusion, I have much pleasure in offering, as a volume, what was originally intended only to have had an ephemeral existence in the columns of a newspaper.

That its perusal may secure it, in its permanent form, a liberal share of public patronage, and favourable consideration, is the hope of

JAMES CLARK,
Late Sergeant Royal Scots Fusiliers.

MAIN STREET, NEWTON-UPON-AYR,
30*th September* 1885.

CONTENTS.

APPENDIX.

COLOURED PLATES.

THE COLOURS OF THE ROYAL SCOTS FUSILIERS.

UNIFORM OF THE ROYAL SCOTS FUSILIERS IN 1742.

UNIFORM OF THE ROYAL SCOTS FUSILIERS IN 1827.

UNIFORM OF THE ROYAL SCOTS FUSILIERS IN 1847.

UNIFORM OF THE ROYAL SCOTS FUSILIERS IN 1860.

UNIFORM OF THE ROYAL SCOTS FUSILIERS IN 1885.

THE TWENTY-FIRST REGIMENT,

OR THE

Royal North British Regiment,

NOW THE

Royal Scots Fusiliers.

CONTENTS

OF THE

HISTORICAL RECORD AND REGIMENTAL MEMOIR.

YEAR

1692. Battle of *Steenkirk*.

1693. Battle of *Landen*.

1694. Performed many marches, and arrived at *Deinse*.

Directed to take rank and precedence as the *Twenty-first* Regiment of Infantry.

1695. Surrender of the Town of *Deinse* by Brigadier-General O'Farrell.

Appointment of Colonel Robert Mackay, in place of Brigadier O'Farrell, cashiered by a General Court-Martial.

1696. Proceeded to the Camp at Marykirk, and served with the Army of Brabant.

1697. Appointment of Colonel Archibald Row to the Colonelcy, in succession to Colonel R. Mackay, deceased.

Treaty of Peace concluded at Ryswick.

Regiment returned to Scotland.

1702. Death of King William III.

Accession of Queen Anne.

Declaration of War with France and Spain.

Embarked from Scotland for Holland.

1703. Joined the Allied Army at Maestricht.

Siege and Capture of *Huy*.

Siege and Capture of *Limburg*.

1704. Marched from Holland into Germany.

Engaged in the Battle of *Schellenberg*.

Engaged in the Battle of *Blenheim*.

The three Field-Officers, Brigadier-General Row, Lieutenant-Colonel Dalyel, and Major Campbell, killed in obtaining the glorious victory of Blenheim.

Appointment of John, Viscount Mordaunt, to the Colonelcy, in succession to Brigadier-General Row.

1705. Completed with recruits from Scotland, and engaged in forcing the French Lines at *Helixem* and *Neer Hespen*.

1706. Engaged in the Battle of *Ramillies*.

Engaged in the Capture of *Ostend*, *Menin*, and *Aeth*.

Appointment of Colonel Sampson de Lalo, from the Twenty-eighth Regiment, in exchange with Viscount Mordaunt.

1707. Engaged in Marches, &c., in West Flanders.

The Union of Scotland and England took place, and certain Additions and Alterations were made in consequence in the Colours and Titles of Regiments.

YEAR

1738. Appointment of Colonel John Campbell to the Colonelcy, in succession to Sir James Wood, deceased.

1739. War declared against Spain.

1740. Removed from Ireland to South Britain.

1741. Encamped on Lexden Heath.

1742. Embarked for Flanders.

1743. Marched for Germany, and engaged at the Battle of *Dettingen*.

1744. Encamped at Asche and Alost.

Returned to Ghent for Winter Quarters.

1745. Marched to the Relief of Tournay.

Engaged at the Battle of *Fontenoy*.

Placed in Garrison at Ostend.

Charles Edward, eldest son of the Pretender, landed in Scotland.

Regiment ordered to return from Flanders.

1746. Proceeded to Scotland, and engaged at the Battle of *Culloden*.

Removed to Glasgow.

1747. Re-embarked for the Netherlands.

Engaged at the Battle of Val.

1748. Treaty of Peace concluded at Aix-la-Chapelle.

Returned to England.

1751. Regulations, prescribed by Royal Warrant, for establishing Uniformity in the Clothing, Standards, and Colours of Regiments, &c. &c.

Received the Commendations of the Duke of Cumberland, for Good Conduct in Quarters and Bravery in the Field.

Embarked for Gibraltar.

1752. Appointment of the Earl of Panmure to the Colonelcy, in succession to Lieutenant-General Campbell, removed to the Second Dragoons, Scots Greys.

1760. Returned from Gibraltar to England.

1761. Embarked on an Expedition to Belleisle.

Capture of the Island.

Returned to England.

1763. Proceeded to Scotland.

1765. Embarked for America, and quartered in West Florida.

1770. Removed to Canada.

Appointment of Major-General Hon. Alexander Mackay to the Colonelcy, in succession to Lieutenant-General the Earl of Panmure, removed to the Scots Greys.

YEAR

1772. Returned from Canada to England.

1775. Commencement of the American War of Independence.

1776. Re-embarked for America, and engaged in the Relief of Quebec.

1777. Engaged in Active Operations on Lake Champlain, at Ticonderago, and other places, with the American Troops.

The British Troops, under Lieutenant-General Burgoyne, surrendered.

1781. Returned to Europe, and stationed in Scotland.

1783. Removed to Ireland.

1789. Embarked for Nova Scotia.

Appointment of General Hon. James Murray, from the Thirteenth Regiment, to the Colonelcy, in succession to Lieutenant-General Hon. A. Mackay, deceased.

1793. Embarked for the West Indies.

Proceeded to Martinique to aid the French Royalists.

1794. Engaged in the Capture of Martinique, St Lucia, and Guadaloupe.

Guadaloupe recaptured by the French.

Appointment of Major-General James Hamilton to the Colonelcy, in succession to General Hon. James Murray, deceased.

1796. Returned from the West Indies.

Proceeded to Scotland to recruit.

1800. Embarked for Ireland, after completing its numbers.

1802. Received the Compliments of the principal Inhabitants of Enniskillen for its Excellent Conduct.

Establishment reduced in consequence of the Peace with France concluded at *Amiens*.

1803. Removed to Dublin.

The Establishment again augmented in consequence of a Renewal of War with France.

Insurrection at Dublin.

The Lord Chief-Justice, Viscount Kilwarden, murdered by Populace ; his nephew, the Rev. R. Wolfe, wounded ; and many other Acts of Violence committed.

Regiment assembled to suppress the Riots, and Lieutenant-Colonel Brown murdered by the Insurgents on his way to the Station of the Regiment.

The Command of the Regiment devolved on Major Robertson.

b

YEAR

1803. Received the Approbation and Thanks of the Commander-in-Chief, and of the Civil Authorities in Dublin, for the Exertions used in restoring Tranquillity.

Appointment of General Hon. William Gordon, from Seventy-first Regiment, to the Colonelcy, in succession to General Hamilton, deceased.

1804. Measures adopted for Repelling the Threatened Invasion of the French.

A Second Battalion added to the Regiment, composed of Men raised under the "*Additional Force Act*" in the Counties of Ayr and Renfrew.

1805. First Battalion embarked from Ireland for Portsmouth.

Removed to Weymouth, and reviewed by His Majesty King George III., and other Members of the Royal Family.

Removed to Lewes.

1806. Marched to London to attend the Funeral of Admiral Viscount Nelson, who was killed at the Battle of Trafalgar, and was honoured with a Public Funeral at St Paul's Cathedral.

First Battalion embarked for Sicily.

Second Battalion embarked from Scotland for Ireland.

1807. Hostilities with the Grand Seignior.

First Battalion embarked from Sicily on an Expedition to Egypt; landed at Alexandria, and marched to Aboukir.

Peace with the Turks being restored, the Battalion returned to Sicily.

1809. Flank Companies engaged in the Capture of the Islands of *Ischia* and *Procida*, in the Gulf of Naples.

Attempt made to reduce the Castle of *Scylla*.

The Invasion of Sicily by Murat, King of Naples, defeated.

1811. Second Battalion embarked from Ireland for Scotland.

1814. First Battalion embarked for Italy with a Force under Lieutenant-General Lord William Bentinck.

Landed at *Leghorn*, marched to *Pisa*, thence to Lucca.

Advanced to *Genoa*, and took Possession of that Town and Fortress.

Second Battalion embarked from Scotland for Holland.

Employed in the Attack of Bergen-op-Zoom.

Hostilities on the Continent ceased.

Abdication of Napoleon Bonaparte.

YEAR

1814. Second Battalion embarked for England, and returned to Scotland.

First Battalion embarked for Service in America.

Defeated the American army at *Bladensburg.*

Advanced to *Washington*, captured the City, and destroyed the Arsenal, Docks, &c.

Marched back to St Benedict.

Re-embarked and landed at North Point.

Advanced towards *Baltimore*, and engaged the American Troops.

Major-General Robert Ross killed, and the Command devolved on Colonel Brooke, Forty-fourth Regiment.

Attacked and defeated the American Army at *Godly Wood.*

Colonel Paterson, Twenty-first Regiment, commanded a Brigade, and commended in the Public Despatches.

Attack on the Town of Baltimore abandoned, and the British Troops re-embarked on board the Fleet.

Proceeded to Jamaica, and there reinforced by a strong Detachment from the Second Battalion.

Re-embarked, and proceeded to make an Attack on *New Orleans.*

1815. Major-General Hon. Sir Edward Pakenham killed, and many other Officers and Soldiers killed, wounded, or made prisoners.

The Capture of New Orleans abandoned.

Capture of Fort Bowyer.

Peace with America concluded.

First Battalion returned to the West Indies, and thence to Portsmouth, and proceeded to Cork.

Napoleon Bonaparte returned to France, and regained Possession of that Kingdom.

The Battle of Waterloo took place.

First Battalion embarked from Ireland for Ostend, and proceeded to join the Army under the Command of Field-Marshal the Duke of Wellington.

Formed Part of the Army of Occupation in France.

1816. Second Battalion disbanded at Stirling.

Reviewed by Field-Marshal the Duke of Wellington.

Appointment of Lieutenant-General Lord Forbes, from Fifty-fourth Regiment, to the Colonelcy, in succession to General Hon. William Gordon, deceased.

1817. Proceeded to Calais, and embarked for England.

YEAR

1818. Marched to Portsmouth.

Officers authorised to wear long coats.

1819. Embarked for the West Indies.

Received the particular Thanks of Major-General Lord Howard of Effingham, commanding at Portsmouth, for its Excellent Qualities.

Landed at Barbadoes, and inspected by Lieutenant-General Lord Combermere.

1821. Proceeded to Demerara.

Lieutenant-Colonel J. M. Nooth died, and succeeded by Lieutenant-Colonel J. Leahy.

1823. Insurrection among the Negroes at Demerara.

Received the Thanks of the Lieutenant-General Commanding in the West Indies, of His Royal Highness the Duke of York, and of His Majesty King George IV., for its Conduct in suppressing this Revolt.

Certain Sums voted by the Court of Policy of Demerara to the Regiment, for its Efficient Services on this Occasion.

1824. Removed to St Vincent and Grenada.

1827. Embarked for England.

Arrived at the Isle of Wight, marched to Windsor, and furnished the Duties at the Castle.

1828. Removed from Windsor to Portsmouth.

Marched to Bath and thence to Bristol.

Embarked for Ireland.

1831. Marched to Dublin, and embarked for England.

1832. Removed to Chatham.

1833. Embarked for New South Wales, by Detachments, as Guards over Convicts.

1839. Embarked from Hobart Town for the East Indies.

Arrived at Calcutta.

1840. Removed to Dinapore.

1843. Marched to Kamptee.

1846. Removed to Agra.

1847. Removed to Cawnpore, thence to Calcutta.

1848. Embarked for England, and arrived at Gravesend.

Marched to Canterbury.

Proceeded to Edinburgh.

Retirement of Lieutenant-Colonel G. Deare.

YEAR
1881. Arrival at Dover.
1882. "Blenheim," "Ramillies," Oudenarde," "Malplaquet," and "Dettingen," authorised to be placed on the Colours and Appointments.
Retirement of Major Grahame.
1883. Removal to Aldershot.
1885. Removal to Portland.

THE SECOND BATTALION.

1858. Formation of the Battalion at Paisley.
Removal to Newport in Wales.
1859. Removal to Aldershot.
1861. Removal to Dover.
1863. Embarks for the East Indies, and stationed at Bellary.
1866. Removed to Secunderabad.
1868. Removed to Burmah.
1872. Removed to Madras.
1873. Embarked for England, and stationed on Arrival at Stirling Castle.
1874. Removed to Glasgow.
1875. Removed to Portsmouth.
1877. Removed to Scotland, and quartered at Fort-George.
1878. Removed to Ireland, and quartered at Dublin.
Removed to the Curragh.
1879. Embarked for South Africa.
Engaged in Active Operations and Marches.
Battle of Ulundi.
Retirement of Colonel Collingwood.
Destruction of Sekukuni's Stronghold.
1880. Death of Major Hazlerigg.
Grant of South African Medals.
1881. Defence of Potchefstroom.
Defence of Pretoria.
Protecting and Returning Colours of Ninety-fourth Regiment.
1882. Embarked for the East Indies, and on Arrival stationed at Secunderabad.
1884. Removal to Burmah.

Introduction.

PRIOR to the Restoration all the forces of horse and foot raised in Scotland were purely feudal levies. The armies raised to oppose King Charles I., and those afterwards raised in support of his son, come under this category.

In the time of the Civil War the proportion of horse and foot was allocated upon the counties and burghs in certain defined proportions, and the levies were placed under the command of noblemen from whose lands they were raised, or who possessed the greatest amount of influence in the district, *the lieutenants, colonels, and majors being officers who had served in Scots regiments abroad.*

After the Restoration, although the feudal system would seem at first sight to have been discarded in the process of forming a standing army, the alteration, so far as Scotland was concerned, was more apparent than real, those who were commissioned to raise troops or regiments doing so as exclusively as possible from their own relatives and vassals.

The earliest list of the Scots establishment extant is that dated 1678. It is as follows:—

1. HIS MAJESTIES GUARD OF HORSE.
2. HIS MAJESTIES REGIMENT OF FOOTE GUARD.
3. THE GARRISON OF EDINBURGH CASTLE.
4. THE GARRISON OF STRIVELING (Stirling) CASTLE.
5. THE GARRISON OF DUMBARTON CASTLE.
6. THE GARRISON OF THE BASS.
7. THE FOOTE REGIMENT COMMANDED BY THE EARLE OF MARR.

8. THE FOOTE COMPANY OF HIGHLANDMEN.
9. THE TROOP OF HORSE COMMANDED BY THE EARLE OF HOME.
10. THE COMPANY OF DRAGOONS COMMANDED BY THE VISCOUNT OF KINGSTON.

By His Majesties command,
LAUDERDALE.*

WHITEHALL, *19th October 1678.*

Many alterations were made in the above forces during the reign of Charles II., from 1660 to 1685, but "*The Foote Regiment commanded by the Earle of Marr*" appears to have remained intact, and amidst the various changes is still borne in the Army List as "*The Royal Scots Fusiliers,*" having been formerly known as the "*Twenty-First Royal North British Fusiliers,*" and in the days of old enjoyed the *sobriquet* of "*The Earle of Marr's Grey Breeks,*" probably from the title of the nobleman who raised the regiment, and from the colour of its uniform, or more correctly the nether portion of it, although it is still open to question and antiquarian research to substantiate the impression that the "Scots Army" of those early days were clothed in a complete uniform of a *grey* colour.

After the Restoration the comparatively small bodies of men who were intended to form the nucleus of a standing army were raised by one unvarying method, *i.e.*, by independent troops and companies, which in several cases were many years in existence before they were regimented; to this arrangement "*The Earle of Marr's Foote Regiment*" was an exception, it having been formed into a regiment in the year it was originally raised from independent companies.

* From Ross's *The Old Scottish Regimental Colours*, p. 9.

HISTORICAL RECORD

AND

REGIMENTAL MEMOIR

OF

The Royal Scots Fusiliers

FORMERLY

THE 21st ROYAL NORTH BRITISH FUSILIERS.

Part I.

1678. The Royal North British Fusiliers derives its origin from the commotions in Scotland during the reign of King Charles II., who attempted to establish Episcopacy in that country, but was opposed by the Presbyterians, who wished to adhere to the religious institutions of their forefathers; and prosecutions being used in Scotland by the Government to enforce obedience, collisions occurred between the inhabitants and the military, which were sometimes attended with loss of life. Several Highland clans were called out in 1678, and quartered upon the Presbyterians, and in the autumn of the same year two regiments were raised from the troops then serving as independent companies, one of which was *the Earl of Mar's regiment* (now the Royal Scots Fusiliers), with the following establishment and rates of pay:—

"*The Foote-Regiment commanded by the Earl of Marr.*

Colonel, as Colonel, . . .	12s. per diem.	
Lieutenant-Colonel, as such, .	7s.	,,
Major, as Major, . . .	5s.	,,
Quartermaster, ...	4s.	,,
Chyrurgeon and Mate, . .	5s.	,,
Marshall, ...	2s.	,,

Eight companies of foote belonging to that regiment, and to each company thereof:—

Captain, as such, . . .	8s. od. per diem.		
Lieutenant, ...	4s. od.	,,	
Ensigne, ...	3s. od.	,,	
2 Sergeants, each, . .	1s. 6d.	,,	*inde.*
3 Corporals, each, . .	1s. od.	,,	*inde.*
2 Drummers each, . .	1s. od.	,,	*inde.*
100 Souldiers, each, . .	os. 5d. Scots."		

The financial duties now performed by the *Army Pay Department* appear to have been combined with those of the quartermaster. By reference to ancient pay-lists, muster rolls, and Treasury payments, it will be seen that sums on account of the pay of the Earl of Mar's regiment were made on several occasions to *Major Wood, quartermaster of the Earl of Mar's foote regiment and his successors.*

The commission of Charles the fifth Earl of Mar, as colonel of the regiment, is dated 23d September 1678, and the corps appears on the records under the establishment for that year.

Most regiments at this period were armed with pikes and muskets, but the practice was introduced of arming every man of a few select corps with a fusil or light musket, and these regiments were called *Fusiliers.* The Earl of Mar's regiment was one of the first which obtained this distinction; the exact date at which the regiment was so armed does not appear. It is also on record that the Earl of Mar's regiment in 1677 was ordered to form a *Grenadier Company, to be instructed in all things belonging to artillery, as gunnery, casting of hand grenadoes, and fyreworks;* and for which purpose a special equipment was issued to said company.

1679. During this year the regiment was engaged in repressing an insurrection of the Covenanters; and under the command of James, Duke of Monmouth, took part in the battle of *Bothwell Bridge*.

1685. The Duke of Argyle having raised the standard of rebellion in Scotland against James II., the regiment formed part of the royal forces commanded by the Earl of Dumbarton. The opposing forces encamped in sight of each other on the 19th of June. The rebels attempted to avoid an engagement by a night march, but were led into a bog in the dark; alarm and disorder ensued, the insurgents flying in various directions, and so left the Earl of Argyle without an army. After the termination of this service the Fusiliers occupied winter quarters in Scotland, being distributed as follows: six companies at Ayr, three at Glasgow, two at Paisley, and one at Inverness,—twelve companies in all, being the establishment at this date.

1686. The Earl of Mar was succeeded in the colonelcy of the regiment by Colonel Thomas Buchan, from a regiment of horse.

1688. When the Prince of Orange was preparing an armament for the invasion of England, the Fusiliers was one of the corps which marched from Scotland to support the authority of King James, and in the early part of November, arrived in the vicinity of London. The Prince of Orange landed on the 5th of November, when King James fled to France. *Colonel Buchan's regiment* was ordered by the Prince of Orange to occupy quarters at Witney in Oxfordshire.

In this year two companies were added to the establishment of the regiment, and pay at the rate of 2s. 4d. per diem first allowed for a "drum major."

1689. The Prince and Princess of Orange were elevated to the throne by the titles of King William and Queen Mary.

Colonel Buchan having adhered to the interests of King James, King William conferred the colonelcy of the regiment on Colonel Francis Fergus O'Farrell, by commission dated 1st March 1689.

From Oxfordshire the regiment marched to Gravesend, where it embarked for Holland, joining the Dutch army commanded by Prince Waldeck; served in the campaign of the year with the division under the Earl (afterwards Duke) of Marlborough. The Fusiliers took part in a sharp action with the French troops commanded by Marshal d'Humières at *Walcourt*, in the province of Namur, on the 25th of August, on which occasion the French were repulsed in their attacks on the Allied Army with considerable loss.

1691. In March the regiment was encamped at Halle, in South Brabant, and formed in brigade with the second battalion of the Royals. The French besieged Mons, but the Allies were too few in numbers to prevent the capture of the place by the enemy. After the surrender of Mons, the Fusiliers were placed in quarters until May, when they encamped near Brussels and formed in brigade with the Royals, and the Scots regiments of Mackay, Ramsay, and Angus, under the command of Brigadier-General Ramsay.

In a list of the army in Flanders, printed in July 1691, the regiment is styled O'FARRELL'S FUSILIERS, and its uniform stated to be *red, faced and lined with the same colour.*

At the termination of the campaign, it was again placed in winter quarters.

1692. A numerous French army appeared in the Netherlands in the spring, and besieged Namur, when O'FARRELL'S FUSILIERS were called from their quarters, and advanced with the army, commanded by King William III., to the relief of the place; but the march having been delayed by heavy rains, the garrison surrendered on the 20th of June. A few days afterwards a detachment of the regiment was employed in an attempt to surprise Mons, but the garrison was found prepared. On this occasion Colonel Sir Robert Douglas and Colonel O'Farrell, having proceeded a short distance to consult with the Prince of Wirtemberg, who commanded the party, mistook their way in the dark, and were made prisoners by a detachment of French cavalry. They were released on paying the regulated ransom.

The FUSILIERS formed part of the advanced guard at the battle of *Steenkirk*, on the 3d of August, and were severely engaged with superior numbers of the enemy under the Duke of Luxembourg. The regiment distinguished itself on this occasion, and sustained the loss of many brave officers and soldiers. D'Auvergne states, in his history of the campaign,—"Our vanguard behaved in this engagement to such wonder and admiration that though they received the charge of several battalions of the enemy, one after another, yet they made them retreat almost to their camp." The corps in advance were not supported in time to enable them to persevere in their career of victory, and King William commanded the army to retire.

In this affair the following casualties occurred amongst the officers:—Killed, Captains White, Cygnoe, Mackenzie, and Sharp, Lieutenants Charles King and Edward Griffith; wounded, Lieutenant Newton.

1693. Taking the field in summer, the North British Fusiliers were formed in brigade with the regiments of Leven (Twenty-fifth), Munro (Twenty-sixth), Mackay and Lauder (afterwards disbanded), under the command of Brigadier-General Ramsay, and after taking part in several manœuvres were engaged at the battle of *Landen* on the 29th of July. At sunrise on the morning of that day a French force of greatly superior numbers, commanded by the Duke of Luxembourg, appeared before the position occupied by the Confederate Army, under King William III., when the Fusiliers, and other regiments of their brigade, were ordered to occupy some hedges and narrow roads beyond the village of *Laer*, on the right of the line. This village and ground, occupied by General Ramsay's brigade, being attacked by a numerous body of the enemy, the North British Fusiliers were engaged in a sharp musketry battle in the open ground. At length the Third Foot, and other corps in the village of

Laer, were forced to retire, but they rallied, and, being joined by Brigadier-General Ramsay's brigade, the whole charged, and by a gallant effort recaptured the village. The regiment distinguished itself on this occasion. The French afterwards carried the village of *Neer-Winden* and the position. The regiments at Laer then became separated from the main body of the Confederate Army; they gallantly defended their post for some time, and eventually retired fighting to the Gheet, forded that river, and joined several corps which had crossed the bridge of Neer-Hespen. The army was retreating, and the Fusiliers accompanied King William to the vicinity of Tirlemont. The regiment had Captains Campbell and Strayton, Lieutenants Douglas and Dunbar, and Adjutant Walle wounded, Captain Paterson taken prisoner, and a number of soldiers killed, wounded, and prisoners.

At the end of the campaign the regiment was placed in garrison at Bruges.

1694. During the summer of this year the regiment performed many long marches in Brabant and Flanders, and in the autumn it marched to Deinse. This year the King commanded a board of general officers to assemble, and decide upon the rank of the several corps of the army. This board gave precedence to the English regiments, and gave the Scots and Irish regiments rank in the English army from the date of their first arrival in England, or from the date when they were first placed on the English establishment. The NORTH BRITISH FUSILIERS, not having entered England until the Revolution in 1688, received rank as TWENTY-FIRST Regiment. Numerical titles were not generally used until the reign of George II.

1695. When the army took the field to serve in this year's campaign the Twenty-first were left in garrison at Deinse, where some stores of provisions were formed. King William undertook the siege of Namur, and the regiment was directed to join the covering army under the Prince of Vaudemont; but it subsequently returned to Deinse, of which place its colonel, Brigadier-General O'Farrell, was commandant.

The French commander, Marshal Villeroy, detached a strong body of troops, under the Marquis of Feuqueres, to reduce the town of Deinse, where the *Fusiliers* were stationed. This town was situated on the river Lys ; it was only slightly fortified, and in many places there was only an entrenchment and some palisades as defensive works. Eight pieces of cannon were the only ordnance in the town. Under these circumstances, Brigadier-General O'Farrell considered it impossible to make a successful defence of the place, and he surrendered on the 21st of July without having fired a shot. The Fusiliers became prisoners of war on this occasion.

Brigadier-General O'Farrell was afterwards tried by a court-martial, and cashiered, and King William conferred the colonelcy of the regiment on Colonel Robert Mackay, from a Scots corps.

After the surrender of *Namur* to the Confederate Army, the regiment rejoined the Allied forces, and was again stationed at Bruges.

1696. From Bruges the regiment proceeded to the camp at Marykirk, and it served in the campaign of this year with the army of Brabant ; in the autumn it went into village cantonments. In December, Colonel Robert Mackay died.

1697. On the 1st January, King William conferred the colonelcy of the regiment on Lieutenant-Colonel Archibald Row, from the Sixteenth Foot.

Quitting their village quarters on the 13th of March, the Fusiliers entered upon the operations of another campaign. While the troops were in the field, negotiations for a general peace commenced at Ryswick, and the treaty was signed in September.

The regiment returned to Scotland in the winter, and was stationed there during the remainder of King William's reign.

While stationed in Scotland the FUSILIERS were most popular, and had no difficulty in filling their ranks with a superior class of recruits, in addition to providing drafts for other Scotch regiments serving in Flanders.

1702. Queen Anne succeeded to the throne on the 8th of March; and the French monarch having violated the conditions of existing treaties by procuring the accession of his grandson, Philip, Duke of Anjou, to the throne of Spain, war was declared against France soon afterwards. At the commencement of hostilities the TWENTY-FIRST was selected to proceed on foreign service, and embarked from Scotland for Holland to serve with the Allied Army commanded by the Earl of Marlborough. The regiment did not join the army immediately on its arrival in Holland, but was stationed some time at Breda, and in September it marched towards Flanders.

1703. Leaving its winter quarters in April, the regiment marched to Maestricht, where the Allied Army was assembled, and with the second battalion of the Royals, Tenth, Sixteenth, and Twenty-sixth regiments, was formed into a brigade under Brigadier-General the Earl of Derby.

The TWENTY-FIRST took part in the operations of the campaign, and its services were connected with the reduction of *Huy*, a strong fortress on the Maese, above the city of Liege, which was besieged and captured in ten days. It was afterwards detached from the main army to take part in the capture of *Limburg*, a city of the Spanish Netherlands, situated on a pleasant eminence on the banks of the Wesdet. The siege of this place was commenced on the 10th of September; the FUSILIERS were employed in carrying on the approaches, and in making the attacks; in seventeen days the garrison surrendered at discretion. In October, the regiment marched back to Holland, where it was stationed during the winter.

1704. From Holland the Fusiliers marched, in the months of May and June, to the interior of Germany, to arrest the progress of the French and Bavarians, who had gained considerable advantage over the Imperialists. A junction was formed with the Germans, under the Margrave of Baden; and on the 2d of July the TWENTY-FIRST took part in the attack of the enemy's position on the lofty heights of *Schellenburg*, on the north bank of the Danube, when the entrenchments were carried, and the French and Bavarians, commanded by the Count d'Arco, were driven from their post with severe loss.

The regiment had a few private soldiers killed and wounded; also, Captain Kygoe, Lieutenant Johnson, and John Campbell, wounded.

After this victory, the North British Fusiliers penetrated the Electorate of Bavaria to the vicinity of the enemy's fortified camp at Augsburg, which was found too strong to be attacked with any prospect of success, and the army retired in order to undertake the siege of Ingoldstadt. At the same time a numerous reinforcement of French troops arrived at the theatre of war.

These events were followed by the battle of *Blenheim* on the 13th of August, when the French and Bavarians, commanded by Marshal Tallard and the Elector of Bavaria, were overpowered by the Allies under the Duke of Marlborough and Prince Eugene of Savoy, and a victory was gained which reflected lustre on the British arms. The FUSILIERS were selected to lead the attack against the French troops in the village of Blenheim, and their colonel, Brigadier-General Row, placed himself at the head of his regiment, which was followed by four other corps. In the annals of Queen Anne it is stated:—"The five English battalions, led on by Brigadier-General Row, who charged on foot at the head of his own regiment with unparalleled intrepidity, assaulted the village of Blenheim, advancing to the very muzzles of the enemy's muskets, and some of the officers exchanged thrusts of swords through the palisades;" but the avenues of the village were found strongly fortified and defended by a force of superior numbers. Brigadier-General Row led the North British Fusiliers up to the palisades before he gave the word "*fire*," and the next moment he fell mortally wounded. Lieutenant-Colonel Dalyel and Major Campbell, being both on the spot, stepped forward to raise their colonel, and were both instantly pierced by musket balls; the soldiers, exasperated at seeing their three field officers fall, made a gallant effort to force their way into the village, but this was found impossible, and they were ordered to retire. The moment the Fusiliers faced about, thirteen squadrons of French cavalry galloped forward to charge them, and one of the colours of the regiment was captured by the enemy; but the French horsemen were repulsed by the fire of a brigade of Hessians, and the colour recovered.

Another attempt to capture the village of Blenheim having failed, the firing was continued against this post, and the army advanced against the enemy, who was driven from the field with great slaughter, the loss of its cannon, and many prisoners, among whom was the French commander, Marshal Tallard. Additional forces were then brought against the French troops in Blenheim, and they surrendered prisoners of war. The Germans, who attacked the enemy's right, were also victorious; and the gallant achievements of the Allied Army raised on the banks of the Danube, a trophy which time cannot destroy.

The Fusiliers had Lieutenant-Colonel Dalyel; Captain Stratton, senior; Captain Stratton, junior; Lieutenants Vandergracht, Hill, Campbell, and Trevallion, killed; Brigadier-General Row and Major Campbell died of their wounds; Captains Craufurd and Fairlee, Lieutenants Dunbar, J. Douglas, Elliott, Ogilvy, Maxwell, Stuart, Primrose, and Gordon, wounded.

The prisoners captured were so numerous that the Fusiliers, with four other regiments, were sent to Holland in charge.

The coloneley of the regiment was conferred on John Viscount Mordaunt from the FOOT GUARDS.

1705. The regiment was employed in the expedition up the Moselle; and returning to the Netherlands, was afterwards engaged in forcing the French lines at *Helixem* and *Neer-Hespen*, on the morning of the 18th July, when the superior tactics of the British Commander, and the gallantry of his troops, were very conspicuous.

1706. The NORTH BRITISH FUSILIERS also took part in gaining another splendid victory over the combined

French, Spanish, and Bavarian forces at *Ramillies*, on Whitsunday, 23d of May 1706. During the early part of the action the Fusiliers, the Third Foot, and three regiments of cavalry, were stationed on the heights of Foulz, where they had a view of the field of battle.

An important crisis in the battle arriving, these corps descended from the heights—the Fusiliers and Third Foot forced their way through a morass, crossed the Little Gheet, ascended the acclivity between that river and the Jauche, and charging the enemy's left flank, drove three French regiments into the low grounds, when the greater part of them were either killed or taken prisoners. The Allies were successful on every part of the field, and the legions of the enemy were overpowered, and pursued from the plains of Ramillies with great slaughter until the following morning, by which time nearly all the enemy's cannon, with many standards, colours, and kettle-drums were captured. This victory raised the reputation of the British arms, and was followed by very important results. Spanish Brabant, and many of the principal towns of Flanders, were rescued from the power of the enemy. The services of the regiment are connected with the capture of *Ostend*, *Menin*, and *Aeth;* and it passed the winter in garrison in Flanders. In June of this year, Viscount Mordaunt exchanged with Colonel Samson de Lalo from the Twenty-eighth regiment.

1707. During the campaign of 1707, the services of the regiment were limited to marches and occupying positions; and it passed the winter in West Flanders.

The Union of Scotland and England took place this year, which occasioned St George's Cross to be added to the colours of the Scots regiments, and St Andrew's Cross to the colours of the English regiments. The corps previously designated as *Scots* regiments now took the title of *North British Regiments.*

1708. In May the regiment took the field, and on the 11th of July, participated in the battle of *Oudenarde*, which was fought in the broken grounds near the river Scheldt. The regiment was engaged in a severe musketry action, and it succeeded in driving the French corps opposed to it from field to field, until the darkness of the night put an end to the conflict. Before the following morning the wreck of the French army had retreated in disorder towards Ghent.

After the victory, the siege of *Lisle*, the capital of French Flanders, was resolved upon. This fortress was deemed almost impregnable; it was garrisoned by 15,000 men, commanded by Marshal Boufflers, who resolved upon making a desperate defence. The Fusiliers were selected to take part in the attack on this important fortress under the orders of Prince Eugene of Savoy; the covering army was commanded by the Duke of Marlborough. The regiment had several men killed and wounded in carrying on the approaches, and at the attack of the counterscarp it had thirteen men killed; three officers, four sergeants, and sixty-six rank and file wounded.

The progress of this siege was a subject of peculiar interest throughout Europe; and the besieging army made extraordinary efforts for the capture of the place, which were attended by complete success on the 9th of December, when the citadel surrendered.

1709. The regiment joined the army, and was employed in covering the siege of *Tournay* in July and August. The citadel of Tournay surrendered in the beginning of September, and the army afterwards marched in the direction of Mons.

A numerous French army, commanded by Marshals Villars and Boufflers, took up a position at *Malplaquet*, and strengthened the post by entrenchments and other works of defence. The Duke of Marlborough and Prince Eugene had confidence in the valour and perseverance of the troops under their orders, and they attacked the enemy's formidable position on the 11th of September, on which occasion the heroic valour of the troops was conspicuous; the enemy's entrenchments and *abatis-de-bois* were stormed with distinguished gallantry, the determined resistance of the French was overcome, and another trophy was acquired; but with the loss of many brave officers and soldiers, including the colonel of the Fusiliers, Brigadier-General De Lalo, who was killed at the head of a brigade. In addition to its colonel, the regiment had also Captains Monroe, Wemyss, and Farley, killed ; Captains Montressor and Lowther, wounded.

After the death of Brigadier-General De Lalo, Viscount Mordaunt was reappointed on the 4th of September to the colonelcy of the Fusiliers. The regiment was afterwards employed in covering the siege of *Mons*, which was terminated by the surrender of the garrison on the 20th of October, when the regiment marched into quarters.

1710. On the 14th of April the regiment marched towards the frontiers of France, and was engaged in the movements by which the French lines were pierced at *Pont-a-Vendin;* it was afterwards selected to take part in the siege of *Douay*, where it performed much severe service. It was employed in carrying on the approaches, in storming the outworks and other duties connected with the siege, and sustained considerable loss in killed and wounded. The garrison beat a parley on the 25th of June, and afterwards surrendered the fortress. After the capture of Douay, the regiment was employed in covering the siege of *Bethune*, which place was surrendered in August. The regiment was also with the covering army during the sieges of *St Venant* and *Aire ;* the former place surrendered on the 30th of September, and the latter on 9th of November.

Viscount Mordaunt died this year, and was succeeded in the colonelcy by Major-General Meredith, from the Thirty-seventh regiment. This officer was succeeded in December by Major-General the Earl of Orrery.

1711. After passing the winter in quarters at Dendermond, the regiment joined the army in May. It took part in the movements by which the boasted impregnable French lines were pierced at *Arleux* on the 5th of August. The regiment was afterwards employed in the siege of *Bouchain*, in which service obstacles of the greatest magnitude had to be overcome, and the abilities of the commanders, and the valour of the troops, were put to a severe test ; these qualities were found in the besieging army. On more than one occasion the soldiers fought up to their middle in water, and by a gallant perseverance, which reflected honour on all the corps engaged, every difficulty was overcome, and the garrison surrendered on the 13th of September.

1712. The regiment joined the army commanded by the Duke of Ormond, and advanced to the frontiers of Picardy; but a suspension of hostilities was afterwards proclaimed preparatory to a general peace, when the British army marched to Ghent, and afterwards went into quarters.

1713. A treaty of peace was concluded at Utrecht; and the Fusiliers could look back with exultation at the career of victory and honour which had attended their efforts during these memorable campaigns. At this period the regiment is designated by historians and in the official documents by the title of the ROYAL NORTH BRITISH FUSILIERS, but the date when the honorary distinction of "ROYAL" was conferred upon it has not been ascertained.

1714. The regiment was stationed in Flanders until the decease of Queen Anne, on the 1st of August, and the accession of King George I., when it was ordered to embark for England. It landed at Gravesend on the 23d of August, and was afterwards directed to march to Scotland.

1715. This year the Earl of Mar erected the standard of rebellion in Scotland, and summoned the Highland clans to aid him in placing the Pretender on the throne. The Fusiliers were encamped at Stirling, under the Duke of Argyle, and advanced with the Royal army to *Dunblane*, to defeat the attempts of the enemy to march southward. On the morning of the 13th of November the two armies confronted each other on *Sheriffmuir*. On the approach of the clans, it was found necessary for the Royal forces to change position; whilst doing so, they were suddenly attacked, and suffered severely. The left wing of the rebel army was overpowered, and driven from the field with great slaughter; and the left wing of the Royal army was also forced to retire; thus each commander had one wing victorious, and one wing defeated. The rebels were prevented marching southward and retired; the King's troops returned to their camp at Stirling.

The Fusiliers had one captain, two lieutenants, three sergeants, and eighty-five rank and file killed; one captain, one sergeant, and twenty-five rank and file wounded.

1716. Reinforcements having arrived, the King's troops advanced in January to attack the insurgents, who made a precipitate retreat. The Pretender and several leaders in the rebellion escaped to the Continent, and the clans separated. The rebellion was thus suppressed.

In July the Earl of Orrery was succeeded in the colonelcy of the regiment by Colonel George Macartney.

1717-1727. The Fusiliers were employed on home service during these years; and in 1727 they were held in readiness to embark for Holland, to aid the Dutch in their approaching war with the Emperor of Germany, but the presence of British troops was not required.

In the latter year, Colonel Macartney was removed to the Seventh Horse, being succeeded by Brigadier-General Sir James Wood.

1728-1737. The order for embarking for Holland having been countermanded, the Royal North British Fusiliers proceeded to Ireland, and were placed upon the establishment of that country.

1738. Major-General Wood having died, the vacant colonelcy was conferred upon John Campbell, afterwards Duke of Argyle.

1739-1740. War having been declared against Spain, in the autumn, the Fusiliers, after a station of twelve years in Ireland, were removed to England, landing at Liverpool.

1741. In the summer of this year the regiment was encamped on Lexden Heath, where seven regiments of cavalry and seven of infantry were assembled, and held in readiness for foreign service.

1742. In the summer, King George II. sent 16,000 men to Flanders, to support the interests of the House of Austria against the aggressions of France and Bavaria.

The Fusiliers formed part of this force, and were stationed for some time at Ghent.

1743. Early in this year the regiment commenced its march for Germany, and, after taking part in several movements in the field, had the honour to distinguish itself under the eye of its sovereign, at the battle of *Dettingen*, on the 27th of June, when the French troops, under Marshal Noailles, were driven from the field of battle with great slaughter, and the loss of a number of standards and colours.

The Fusiliers had Lieutenant Yonge, one sergeant, and thirty-five rank and file killed; Lieutenant Levingstone, one sergeant, two drummers, and fifty-three rank and file wounded.

An interesting anecdote is told of the ROYAL NORTH BRITISH FUSILIERS in connection with this engagement. The French Cuirassiers were bearing down upon the regiment, and would doubtless have worked a deal of mischief, had Colonel Sir Andrew Agnew, their commanding officer, not displayed great coolness and courage. Sir Andrew had not time to form square, but he formed them into a lane; and before the Cuirassiers had time to draw up, they found themselves in the midst of a cross fire, from which there was no retreat.

After the battle was over, King George rode up to the colonel, and addressed him as follows,—"I saw the Cuirassiers get in among your men this morning, colonel." To which Sir Andrew replied, in a peculiarly dry, pawky Scottish style,—"*Ou ay, yer Majestee; but they didna get oot again.*"

The following verses were written upon this incident:—

THE TWENTY-FIRST ROYAL NORTH BRITISH FUSILIERS AT DETTINGEN.

Attention! all ye soldier lads, who love the Twenty-first,
And hear one of its gallant deeds in homely rhyme rehearsed.
On many a hard fought field, my lads, its laurels have been won,
And always true are those who wear the number "Twenty-one."

It was when, in the olden time, they served in Germanie,
Against the pride and power of France, and all her chivalry;
Sir Andrew Agnew at their head, they feared no foreign foe,
But sharp and sure the Frenchmen met, and dealt them blow for blow.

The Frenchmen did not care to face old Scotia's Fusiliers;
So, on the field of Dettingen, they launched their Cuirassiers,
To charge the stubborn phalanx of the sturdy Twenty-first,
And drive for ever from the earth the corps by them accurst.

As, from the Alps, the avalanche comes thundering to the vale,
So charged the Cuirassiers that day, but never could prevail
To shake the stout battalion that stemmed their wild career,
And baffled them, and turned them back with many a ringing cheer.

Three times they charged upon the square, as often they rode back
Disordered, to form up again, and yet again attack;
And then Sir Andrew grimly smiled, and from his square withdrew
A section of its bristling front, to let the French ride through.

Amazement took the Frenchmen then, and laughter loud they raised:
"Are Scotia's Fusiliers now led by a commander crazed?"
And swiftly once again they charge, and ride straight thro' the gap,
And then Sir Andrew, *cannily*, enclosed them in the trap.

The section once more fills the gap, and loud Sir Andrew's call:
"Square! Inwards face! your bayonets will do when fails your ball!"
And so it happened on that day, the fairest troops of France
Were hemmed in by the Fusiliers, and captured horse and lance.

The battle din was over, and all was hushed and still,
When the General met Sir Andrew, and thanked him with good-will:
"The French got in among your men to-day on yonder plain."
"Quite true, your Grace," said Sir Andrew, "but they *didna get oot again.*"

The regiment was afterwards encamped near Hanau; in August it crossed the Rhine, and was employed in West Germany; but in the autumn it returned to Flanders.

1744. During the campaign of 1744, the regiment served with the army under Field-Marshal Wade. It was encamped between Asche and Alost, afterwards on the banks of the Scheldt, and subsequently penetrated the French territory to the vicinity of Lisle; but returned to Ghent for winter quarters.

B

1745. Quitting its cantonments in April, the regiment marched with the army commanded by His Royal Highness the Duke of Cumberland, to the relief of *Tournay*, which fortress was besieged by a numerous French army, which took up a position near the village of *Fontenoy.* The enemy had a great superiority of numbers; but the Duke of Cumberland, trusting to the innate bravery of his troops, resolved to hazard a general engagement on the 11th of May, when the Royal North British Fusiliers had their valour and endurance put to a severe test; and they proved themselves not unworthy successors of the gallant officers and soldiers who triumphed at Blenheim and Ramillies, under the great Duke of Marlborough.

Soon after nine o'clock, the British infantry advanced in the face of a heavy fire of grape and musketry, and, by a gallant charge, broke through the French lines; but the Dutch failed to carry the village of Fontenoy; and a brigade under Brigadier-General Ingoldsby did not capture the battery it was intended to attack. The troops which had forced the enemy's centre were thus exposed to a severe cross fire, and were ordered to retire. A second attack was made. British valour and intrepidity were again triumphant; but the failure of the Dutch a second time produced disastrous results, and the British regiments which had carried the enemy's entrenchments and forced the centre were nearly annihilated by a cross fire. The Duke of Cumberland afterwards ordered a retreat, and the army withdrew from the field of battle to Aeth.

The Fusiliers suffered severely on this occasion; Lieutenants Campbell, Houston, and Sergeant of the regiment were killed; Major Colville, Captains Latan, Olivant, and Knatchbull, Lieutenants Maxwell, Colville, Ballenden, Macgaken, and Townsend, wounded; Captain Sandilands, Lieutenant Stuart, and Quartermaster Stewart, prisoners; eleven sergeants, and 259 rank and file killed, wounded, and prisoners. The severe loss which the regiment had sustained occasioned it to be placed in garrison at *Ostend.* This place was besieged by a numerous French force, and the garrison defended their post some time; but the works were not in repair, the stores were defective, and the garrison not sufficiently numerous. Under these circumstances the governor surrendered, on condition that the garrison should join the Allied Army.

While the regiment was in Flanders, Charles Edward, eldest son of the Pretender, arrived in Scotland, and being joined by a number of Highland clans, he made a desperate effort to overturn the existing Government, and establish his father's authority in the kingdom. The Fusiliers were ordered to return to England. They arrived in the river Thames on the 4th of November, and, after landing, marched northward. The efficiency of the regiment was increased by a body of fine recruits enlisted in Scotland.

1746. The regiment arrived at Edinburgh in January, and advanced with the army commanded by the Duke of Cumberland towards Stirling, when the young Pretender raised the siege of Stirling Castle, and made a precipitate retreat. The pursuit was retarded by severe weather, but the army continued its advance when the season permitted, and on the 16th of April encountered the clans on *Culloden Moor*. The regiment was in the front line on this occasion, and took part in repulsing the attacks of the Highlanders, and in driving them from the field of battle with great slaughter. This victory proved decisive, and the rebellion was suppressed. The loss of the regiment was limited to seven private soldiers killed and wounded. It was encamped a short time at Inverness, and afterwards removed to Glasgow.

1747. From Scotland the regiment was again removed to the theatre of war in the Netherlands, where it arrived in time to take part in the operations of the campaign of 1747; and it was engaged at the battle of *Val* on the 2d of July. On this occasion the Allied Army was very inferior in numbers to the enemy, and although the gallantry of the British infantry was very conspicuous throughout the action, the Duke of Cumberland was obliged to order a retreat to Maestricht.

Eight rank and file of the Fusiliers were killed; one sergeant and fifteen rank and file were wounded; and five men missing.

1748. The regiment was again in the field in the summer. Hostilities were terminated by the Treaty of Aix-la-Chapelle, and during the winter the regiment returned to England.

1749-1751. The regiment during these years was stationed in England until the end of 1751, when it embarked for Gibraltar. Prior to departure it received the commendation of His Royal Highness the Duke of Cumberland, on account of its good conduct in quarters, and for its uniform gallantry in the field.

In the Royal Warrant, issued on the 1st of July 1751, for ensuring uniformity in the clothing, standards, and colours of the army, the following directions are given for the TWENTY-FIRST REGIMENT, OR THE ROYAL NORTH BRITISH FUSILIERS: — "IN THE CENTRE OF THEIR COLOURS THE THISTLE WITHIN THE CIRCLE OF ST ANDREW, AND THE CROWN OVER IT; AND IN THE THREE CORNERS OF THE SECOND COLOUR THE KING'S CIPHER AND CROWN. ON THE GRENADIER CAPS THE THISTLE AS ON THE COLOURS, THE WHITE HORSE, AND MOTTO OVER IT, NEC ASPERA TERRENT, ON THE FLAP. ON THE DRUMS AND BELLS OF ARMS THE THISTLE AND CROWN TO BE PAINTED AS ON THE COLOURS, WITH THE RANK OF THE REGIMENT UNDERNEATH."

1752. Lieutenant-General Campbell was removed to the *Scots Greys*, and Colonel the Earl of Panmure from the Twenty-fifth regiment, succeeded to the colonelcy of the Royal North British Fusiliers, by commission dated 29th April.

1753-1760. The regiment remained at Gibraltar until 1760, when it returned to England.

1761. In the meantime another war had commenced between Great Britain and France, and the ROYAL NORTH BRITISH FUSILIERS, mustering 800 men, under the command of Lieutenant-Colonel Edward Maxwell, sailed with the expedition, under Major-General Hodgson, for the attack of the French island in the Bay of Biscay, called *Belle-Isle.* The fleet appeared before the island on the 7th April, but the coast was found like a vast fortress—the little which nature had left undone by rocks and crags, having been supplied by art. A landing was, however, effected on the following day. The TWENTY-FIRST was one of the regiments which leaped on shore, and stormed the works of *Port Andro*, under a heavy fire of cannon and musketry; the works were found too steep to be ascended, and although the officers and soldiers made a gallant effort, one attempting to lift another up, it was found impossible to succeed, and they were ordered to return on board of the fleet. The regiment had three sergeants, one drummer, and eight rank and file killed; eight rank and file wounded; Lieutenants Innis and Ramage, and thirty-five rank and file prisoners; many of the officers and soldiers taken prisoners were severely wounded, and unable to return on board the fleet when the order was given to retire.

A landing was effected on the 22d of April, at a rugged spot near Point Lomaria, where the difficult ascent had occasioned the enemy to be less attentive to that part of the coast; and the troops, under Brigadier-General Lambert, having landed, gained the summit of the rock, and repulsed the attempts of the enemy to dislodge them—capturing three brass field-pieces. The cannon were afterwards landed from the ships, and dragged up the rocks; the lines which covered the town of Palais were captured, and the siege of the citadel commenced. The ROYAL NORTH BRITISH FUSILIERS took part in the siege of the *Citadel of Belle-Isle*, which was prosecuted with so much vigour, that the governor, the Chevalier de St Croix, was forced to surrender on the 7th of June. The capture of the island was thus effected, with the loss of about 1800 men killed and wounded.

1762-1764. After the surrender of the castle of Belle-Isle, the regiment returned to England, where it was stationed; and in 1763 and 1764, it occupied quarters in Scotland.

1765-1769. On the 6th of May 1765, the regiment embarked for America, and was quartered five years in West Florida.

1770-1771. In 1770 it was removed to Canada, and was stationed some time at Quebec. In November of the same year, Lieutenant-General the Earl of Panmure was removed to the Scots Greys, and was succeeded in the colonelcy of the Royal North British Fusiliers by Major-General the Hon. Alexander Mackay, from the Sixty-fifth regiment.

1772-1774. The regiment returned to England, where it remained until 1775.

1775. The American War commenced this year, and during the winter Quebec was besieged by an American force.

1776. In the spring the regiment embarked for the relief of *Quebec*. On the arrival of the British reinforcements, the Americans raised the siege and made a precipitate retreat; they were pursued up the country and driven from all the posts which they occupied in that province. After these services were performed the FUSILIERS were quartered at St John's, where they were stationed during the winter.

1777-1780. The regiment was employed in active operations in the spring of 1777, with the armament commanded by Lieutenant-General Burgoyne; it embarked in boats on Lake Champlain, and sailed to Crown Point, where the troops halted three days, and afterwards proceeded against Ticonderago; but the Americans quitted the fort without hazarding the events of a siege. The regiment returned on board the flotilla, and, sailing along the lake, arrived about three o'clock on the afternoon of the 6th of July, within three miles of Skenesborough, where the Americans had a stockaded fort. The Ninth, Twentieth, and TWENTY-FIRST regiments leaped on shore, and ascended the mountains to get behind the fort and cut off the retreat of the garrison; but the Americans fled precipitately, and escaped with the loss of a few men made prisoners.

On the 8th of July the regiment was detached towards Fort Anne to support the Ninth, which was attacked by very superior numbers. The enemy was repulsed, and retreated towards Fort Edward.

To follow up these advantages proved a difficult undertaking—trees and other obstacles had to be removed, creeks and marshes to be crossed, forty bridges to be constructed; but by great exertion these difficulties were overcome, and on the 30th of July the army arrived at the bank of the Hudson River, which was crossed by a bridge of boats on the 13th and 14th of September; and on the 19th, the army advanced against the Americans in position on an island called *Still Water*, where a severe action was fought. Lieutenant-General Burgoyne stated in his public despatch:—"About three o'clock the action began by a very vigorous attack on the British line, and was continued with great obstinacy until after sunset; the enemy being constantly supplied with fresh troops. The stress lay upon the Twentieth, TWENTY-FIRST, and Sixty-second regiments, most part of which were engaged nearly four hours without intermission. . . . . Just as night closed the enemy gave ground on all sides, and left us completely masters of the field of battle."

Several other actions occurred, and the regiment sustained considerable loss in killed and wounded; among the former were Lieutenants Currie, Mackenzie, Robertson, and Turnbull; and among the latter, Captain Ramsay and Lieutenant Richardson.

The circumstances under which the troops commanded by Lieutenant-General Burgoyne eventually became placed, assumed a desperate character; their numbers were reduced to about 3500 men able to bear arms, they were environed by 16,000 Americans, their retreat cut off, and they were short of provisions. Under these accumulated difficulties they agreed to lay down their arms, on condition of being sent to England, and of not serving again in North America during the war. These conditions were, however, violated by the American Congress, and the troops were detained some time in the provinces.

1781-1782. The regiment having returned home, was stationed in Scotland; and at the termination of the American War, was placed on a reduced establishment.

1783-1789. In 1783 the Fusiliers proceeded to Ireland, where they remained until the spring of 1789, when they embarked from Cork for Nova Scotia, and landing at Halifax, were stationed in the British provinces in North America nearly four years.

Lieutenant-General the Hon. Alexander Mackay died in 1789, and the colonelcy was conferred on General the Hon. James Murray, from the Thirteenth regiment.

1790-1793. While the regiment was in North America, a revolution took place in France, and republican principles were extended to the French West India Islands, where the inhabitants of colour rose in arms against the European settlers, many of whom sought protection from Great Britain. Under these circumstances, the Royal North British Fusiliers were removed to the West Indies in the spring of 1793. The French Royalists of Martinique sent pressing applications for assistance, and Major-General Bruce, commanding the British troops in the West Indies, was induced to proceed with a small force to their aid. The regiment was employed on this service. It landed at Caise de Navire on the 14th of June, the other corps landed on the 16th. About 1100 British and 800 French Royalists advanced to attack the town of St Pierre, but the Royalists were undisciplined; they got into confusion, fired on one another, and so completely disconcerted the plan of attack, that the English General, not having a force sufficiently numerous for the purpose without them, ordered the British troops to return on board the fleet.

1794. General Sir Charles (afterwards Earl) Grey assembled a body of troops at Barbadoes, in January, for the attack of the French islands, and the flank companies of the TWENTY-FIRST were employed on this service. A landing was effected on the island of *Martinique* in the early part of February, and after some sharp fighting, in which these companies had several men killed and wounded, this valuable possession was delivered from the power of the Republicans.

From Martinique, the Grenadiers, under Prince Edward (afterwards Duke of Kent), the light infantry, and three other corps under Major-General Dundas, embarked on the 30th of March for *St Lucia*, where they arrived on the 1st of April, and the conquest of that fine island was completed in three days.

The flank companies of the FUSILIERS were afterwards employed in the reduction of the island of *Guadaloupe.* A determined resistance was made by the enemy; but the island was captured before the end of April. These companies had several men killed and wounded; Captain Macdonald was also wounded.

After the reduction of Guadaloupe, these companies were removed to Antigua.

The loss of so many valuable colonial possessions was not regarded with indifference by the Republican Government of France, and in June a French armament arrived at *Guadaloupe* for the recovery of that island. The negroes and other men of colour flocked to the standard of Republicanism, they were instantly armed and clad in uniforms, the doctrines of liberty and equality were disseminated among this motley crowd, which led to a frightful catalogue of crime and bloodshed. The flank companies of the Twenty-first were called from Antigua, to aid in the defence of Guadaloupe; they arrived on the 10th of June in the "Winchelsea" ship of war, landed on the 19th at Ance Canot, and were engaged in several arduous services, in which Lieutenants Harry Foley Price, Samuel Knollis, and J. S. Colepeper were wounded; also several private soldiers killed and wounded; but the British troops were not sufficiently numerous to contend with the Republican forces.

Lieutenant-Colonel Colin Grahame, of the TWENTY-FIRST, was appointed to the command of the troops in

Basse Terre, and he defended Bevelle Camp until the 6th of October, when he was forced to surrender, his force having become reduced to 125 rank and file fit for duty.

Three companies of the ROYAL NORTH BRITISH FUSILIERS were engaged in the defence of *Fort Matilda*, under Lieutenant-General Prescott, and the garrison made a resolute resistance, until the place became so much injured by the enemy's fire that it was not tenable. The fort was evacuated during the night ot the 10th of December. One company of the Fusiliers occupied the rampart; the light company, under Lieutenant Wm. Paterson, was stationed on the right of the breach; and the third company, under Captain Mackay, was posted along the Gallion River. They thus covered the embarkation of the garrison and stores, and afterwards retired on board the fleet. The three companies were reduced by casualties to one captain, three lieutenants, six sergeants, and ninety-two rank and file. Lieutenant-General Prescott stated in his despatch :—"That during the whole progress of this long and painful siege the officers and men under my charge have conducted themselves in such a manner as to deserve my warmest praise, bearing their hardships with the utmost patience, and performing their duty with alacrity."

General the Hon. James Murray died in this year, and Major-General James Hamilton, from the Fifteenth regiment, succeeded to the colonelcy of the Twenty-first Fusiliers.

1795. In addition to the casualties in action, the regiment also sustained, during its services in the West Indies (and in this year particularly), severe losses from yellow fever.

1796. The Fusiliers, much reduced in numbers, returned to England, landed at Portsmouth, and proceeded to Scotland.

1797-1799. The Royal North British Fusiliers occupied various stations in Scotland, until June 1800, when it embarked from Portpatrick for Ireland, where its numbers were increased to 800 rank and file, by volunteers from the Scots Fencible regiments, then in that country.

1800-1801. In October 1800 the Fusiliers marched to Enniskillen, where they were quartered nearly two years, during which time their numbers were increased to 1000 men by recruits. The good conduct of the regiment during its stay at this place occasioned it to stand very high in the estimation of the inhabitants; and, on its removal, 100 gentlemen and respectable persons sent a memorial to the Commander-in-Chief, requesting that it might be again quartered at Enniskillen, and offered to defray the expense of removal.

1802. On the 15th of July the regiment arrived at Londonderry, where its establishment was reduced, in consequence of the Peace of Amiens having been concluded with France.

1803. The regiment was removed to Dublin in February; its establishment was again augmented in the summer of this year after the renewal of hostilities with France.

An alarming insurrectionary spirit was manifested at Dublin in the summer of this year; and on the evening of the 23d of July an immense number of persons assembled with firearms and pikes; dragged the Lord Chief-Justice, Viscount Kilwarden, out of his carriage, and murdered him; also wounded his nephew, the Rev. Richard Wolfe; and committed numerous other acts of outrage and violence. At this period the regiment was quartered in Cork Street, Thomas Street, and Coombe Barracks, and it quickly assembled to suppress the riots. Lieutenant-Colonel Brown was murdered by the insurgents as he was proceeding from his quarters to head the regiment. The command devolved on Major Robertson, under whose orders the regiment was actively employed in restoring tranquillity, in which service it had twelve men killed and wounded. The regiment received the thanks and approbation of the Commander-in-Chief in Ireland, Lieutenant-General the Hon. H. E. Fox, for its conduct on this occasion; also, the thanks of the city of Dublin. *Lieutenant Douglas, who commanded the light company, and Adjutant Brady, particularly distinguished themselves, and were each presented with a piece of plate by the city of Dublin*, accompanied with the expression of the gratitude and admiration of the citizens for their gallant exertions.

On the decease of General Hamilton in this year, he was succeeded in the colonelcy by General the Hon. William Gordon, from the Seventy-first regiment.

1804. In July the regiment proceeded to Loughrea. Napoleon Bonaparte having made preparations for the invasion of England, his menace was met by a general display of loyalty and patriotism by the British people, who armed to repel the threatened invasion. Among the precautionary measures adopted at this period, an "Additional Force Act" received the royal assent in July. The men raised for limited service under the provisions of this Act, in the counties of Ayr and Renfrew, were added to the ROYAL NORTH BRITISH FUSILIERS, and were formed into a *second battalion*, which was embodied at Ayr, and placed on the establishment of the army on the 25th of December 1804, and was quartered there till August 1806, when it went to Ireland. Its ranks were largely composed of Ayrshire men, and, from its lengthened stay in the county town, the inhabitants became quite attached to it; several of its officers formed matrimonial alliances with some of the leading families in the district. Lieutenant-Colonel Wilson, Major A. Campbell, and Captain M'Haffie, married respectively daughters of Oswald of Auchincruive, Provost Bowie, and Rankine of Drumdow. Captain M'Haffie became a lieutenant-general, and is still represented in Ayr by his daughter, Mrs William Cowan, Wellington Square.

The news of the battle of Trafalgar reached Ayr during divine service on a Sunday forenoon, November 14, 1805; and it was the second battalion of the Royal Scots Fusiliers, under the command of Colonel Adam, that fired the *feu-de-joie* in honour of the event. This was done in the barrack square, on the regiment's return from church, in presence of a large number of townspeople collected at the north bastion of the Fort. It was in the same year that the second battalion of the Twenty-first Fusiliers, and the Ayr and Newton Volunteers, to the number of 1300 men, were reviewed on the Town Green by the General Earl of Moira, who, on the occasion, was made an honorary freeman of the burgh.

Ten months after its departure from Ayr, the Royal North British Fusiliers were inspected by a general officer at Armagh. This was followed by a dinner, at which the General was entertained by the officers of the regiment. When the mess had broken up, an altercation took place between Major Campbell and Captain Boyd (both no doubt a little excited with wine) regarding a word of command which the former had given in the course of that day's inspection, and in which he had been corrected by the General. Major Campbell maintained that he was right, and that the inspecting officer was wrong. Captain Boyd replied that neither was correct, according to the army regulations. The dispute reached its height, when the captain pointedly reiterated that the word of command as quoted by him was right, and that it was wrong as given by the major. Campbell then left the mess-room, and went to his quarters, where, without sitting down, he drank a cup of tea with his wife, and returned to the mess-room just as Boyd was leaving it. Both went into a side room, where they remained about a quarter of an hour. It appears to have been at this point that a duel was agreed upon, for immediately after the interview Major Campbell delivered a box to a brother officer to be kept in safety, should the issue of the duel prove fatal to him. On parting with Campbell, Boyd went into the barrack square, and had not been there many minutes when he received a message to the effect that a gentleman wished to see him in one of the messrooms. Thither he went, and was shown into the apartment where Campbell was waiting. In a few minutes the report of a shot was heard, then another, whereupon a waiter and two officers went in, and found Captain Boyd wounded, and upbraiding his antagonist for having "hurried" him to fight without the presence of "friends." Subsequently, in answer to a question by Campbell, Boyd acknowledged having said he was "ready" before the fatal shot was fired, and at the same time he extended his hand to his unhappy comrade in token of forgiveness. He died on the following day, June 24, 1807. The combatants had faced each other from opposite corners of the room, at a distance of seven paces, and while receiving his antagonist's bullet in his belly, Captain Boyd had directed his fire towards Major Campbell's head, which he narrowly missed. This was seen on a subsequent examination of the room.

Arriving in Ayr, a few days after this unfortunate affair, Campbell lived in concealment for several weeks among his wife's relatives. A warrant having been issued for his apprehension on a charge of murder, he was arrested in a temporary hiding-place in the vicinity of Greenan Castle, and conveyed to Ireland. His trial, which came off at Armagh on the 5th of August, ended in a verdict of guilty, and he was sentenced to be hanged. The verdict proceeded partly on the general illegality of the practice of duelling, and partly on the ground of this particular affair having occurred without the presence of witnesses. The most strenuous efforts to obtain the remission of his sentence were made by the major's friends, in which they were joined by the jury who had convicted him. A temporary respite was granted by the Lord-Lieutenant of Ireland, in the hope that the royal mercy might be extended to the unfortunate criminal. Mrs Campbell, who was present at the trial, set out for Dublin immediately on hearing the verdict. Crossing the Channel in an open boat, she landed at Holyhead, and reached London within twenty-eight hours. Getting access to the Queen at Windsor, she presented a memorial, imploring Her Majesty's intercession in favour of her husband, stating the circumstances of the duel, and detailing his military services. Falling on her knees, she in the most pathetic terms solicited the intercession also of the princesses who stood beside their royal mother. In a personal interview with the Prince of Wales, Mrs Campbell enlisted the sympathy of His Royal Highness, who addressed the Prime Minister on the subject. The King, however, was inexorable, and the temporary respite having expired, the sentence was carried into effect, during Mrs Campbell's absence, on the 24th of August 1807. The unfortunate man, as well as his relatives, were importunate in their request that death should be by shooting, rather than by hanging. But this could not be conceded. It was a rumour of the time that poison had been secretly conveyed into prison, in order that the ignominy of the scaffold might be avoided, but that Major Campbell had on Christian grounds declined to avail himself of the means of self-destruction. When all hope of a pardon had vanished, he showed a dignified resignation and manly firmness. On making his appearance on the scaffold the entire guard, as a tribute of respect, took off their caps, and the major in return saluted them.

His body was brought to Ayr by a small coasting vessel, and landed at the Ratton Quay on the second day after his execution. It was received by a "fatigue party," which, out of sympathy for the fate of a brave but unfortunate officer, had been sent by the commander of the troops, then quartered in the barracks. These soldiers in undress conveyed the corpse to the Old Churchyard—the relatives being also present, accompanied by the provost, magistrates, and other leading men of the town.

It was from the lips of one of the few original members of the second battalion of the Fusiliers, himself an eye-witness of the event, that we learned the sad fate of another officer of that corps. Referring to the capture of New Orleans by the British troops in 1814, our informant said that on the first night after they landed Captain Conran, of the Twenty-first, commanded a detachment, and after being engaged with the enemy there were a number of them taken prisoners. Among these was an officer of the American army, who delivered his sword and surrendered himself prisoner to the captain (Conran) in command, who invited him to come and warm himself at the fire. While in the act of doing so, the captured American pulled a dagger from under his coat, and stabbed his generous captor to the heart.

1805. On the 30th of April, the first battalion embarked from Monkstown for England, landed at Portsmouth, and was subsequently encamped at Weymouth, where several corps were assembled, and was repeatedly reviewed by the King, and other members of the Royal Family, who expressed their high approbation of the ROYAL NORTH BRITISH FUSILIERS on every occasion on which the corps appeared before them. In the autumn the battalion marched to Lewes.

1806. The first battalion marched to London in January to attend the funeral of Vice-Admiral Lord Viscount Nelson. The interment took place on the 9th of January, in St Paul's Cathedral. The battalion afterwards marched to Colchester; and in April embarked from Tilbury for *Sicily*, to protect that island against the French, and landed at *Messina* on the 26th of July.

On the 15th of August *the second battalion of the Fusiliers* embarked from Portpatrick for Ireland, where it was stationed during the following five years.

1807. The Court of the Grand Seignior having become involved in hostilities with Great Britain, the first battalion embarked from Sicily on the 17th of May, and joined the expedition to *Egypt*, under Major-General Alexander Mackenzie Fraser. The battalion landed at Alexandria, and marched to the camp at Aboukir. Peace having been concluded with the Turks, the battalion returned to Sicily, where it arrived in October.

1808. The first battalion occupied quarters in Sicily.

1809. In June, Lieutenant-General Sir John Stuart, commanding-in-chief in the Mediterranean, resolved to menace the capital and kingdom of Naples, as a diversion in favour of the Austrians, who were contending with numerous difficulties in their war with France. The flank companies of the Twenty-first were employed in this service; and, after menacing a considerable part of the coast, which produced much alarm, the romantic and fruitful island of *Ischia*, celebrated for the beauty of its scenery, and situate in the Bay of Naples, about six miles from the coast, was attacked. A landing was effected in the face of a formidable line of batteries, from which the enemy was speedily driven; Lieutenant Cameron of the Twenty-first, who was attached to the British flotilla, attacked the enemy's gunboats with great gallantry and captured twenty-four of their number, but was mortally wounded at the moment of victory. The siege of the castle was undertaken, and in a few days the garrison was forced to surrender. The island of *Procida* surrendered on being summoned. Two valuable islands were thus rescued from the power of General Murat, whom the Emperor Napoleon had nominated King of Naples, in succession to Joseph Bonaparte, upon whom the Emperor had conferred the crown of Spain; and 1500 regular troops, with 100 pieces of ordnance, were captured.

An attempt was, at this period, made to reduce the castle of *Scylla*, but the large force which the enemy possessed in Calabria, rendered this impracticable. The battalion companies of the regiment were employed in this service, and had Captain Hunter killed, eight rank and file wounded.

A detachment of the regiment was sent, at the request of the inhabitants, to the town of Valmi, for the protection of the place; but it was intercepted by the enemy, and Captains Mackay and Conran, Lieutenants M'Nab and Mackay, four sergeants, two drummers, and seventy-six rank and file were made prisoners.

1810. In the summer, General Murat assembled upwards of 100 heavy gunboats, a number of others more lightly armed, and about 400 transport-boats, and brought 30,000 troops to the coast of Calabria, for the purpose of invading Sicily. The ROYAL NORTH BRITISH FUSILIERS were employed on the coast watching the approach of the enemy, and were at the alarm-post, under arms, every morning two hours before daylight for several months. During a dark night, between the 17th and 18th of September, 4000 men, under General Cavaignac, made good their passage, and commenced landing about seven miles to the southward of *Messina*. The alarm being given, the *Twenty-first regiment*, commanded by Lieutenant-Colonel Adam (afterwards General the Right Hon. Sir Frederick Adam, G.C.B., head Colonel of the Twenty-first Royal North British Fusiliers), hurried to the spot, accompanied by two field-pieces, which were attached to the regiment, and prevented several of the boats from reaching the shore. As the boats were retiring, a few of them were sunk by the fire of the field-pieces. The regiment next turned towards that portion of the enemy which had landed, and had taken post on two hills. The "flankers" were thrown out, and a fire of musketry was kept up until daylight, when the enemy, being cut off from the boats and surrounded, surrendered prisoners of war; delivering up one stand of colours. The prisoners, amounting to about 1000 officers and soldiers, were marched to *Messina*. This repulse, with the destruction of so many of the enemy's gunboats by the British and Sicilian flotillas, disconcerted the plans of Murat, and no further attempts were made against Sicily.

1811. In September the second battalion of the Fusiliers embarked from Belfast for Scotland; and in this year sent a strong detachment of volunteers from the militia to join the first battalion in Sicily.

1812. Being still quartered in Sicily, in November the grenadier company of the Fusiliers (first battalion) proceeded to the eastern coast of Spain, to take part in the war of Spanish and Portuguese Independence, under the command of the Duke of Wellington. They arrived at *Alicante* on the 2d of December, but circumstances occurred which occasioned their return to Sicily in the spring of 1813.

1813. Two companies (first battalion) proceeded to the island of *Ponza*.

1814. The brilliant success of the British troops in the *Peninsula*, and of the armies of the Allied Sovereigns on the Continent of Europe, was followed by the embarkation of a body of troops for Italy, under Lieutenant-General

Lord William Bentinck and Major-General H. T. Montressor. The first battalion embarked for this service in February, under Major Whitaker (Colonel Paterson commanding a brigade), and landed at Leghorn on the 13th of March; on the 23d it marched to *Pisa*, and on the 25th to *Lucca*. In April the battalion advanced upon *Genoa;* on the 12th of that month the enemy were driven from *Mount Facia* and *Nervi*, and the British took post at *Sturla*. On the 17th of April, at daybreak, the French position in front of Genoa was attacked; the enemy were driven from the strong position they occupied, and afterwards evacuated the town, which was taken possession of on the 19th of April by the *Twenty-first* and other corps. The regiment had Lieutenant Sabine wounded; one sergeant and fourteen rank and file killed and wounded.

Meanwhile the *second battalion* had been withdrawn from Scotland to take part in the war on the Continent; it embarked from Fort-George, on the 30th of December, landed in Holland on the 10th of January 1814, and was employed in the attack on *Bergen-op-Zoom* on the night of the 8th of March. One portion of the battalion formed part of the third column, under the command of Lieutenant-Colonel Robert Henry, of the Twenty-first, who was directed to draw the enemy's attention to an attack near *Steenbergen gate;* the flank companies were attached to the fourth column, under Brigadier-General Gore. Some severe fighting took place, and advantages were gained in the first instance; but the attack failed, and a number of officers and men who had penetrated the works were forced to surrender prisoners of war. The battalion had a number of men killed and wounded on this occasion; Lieutenant John Butteel died of his wounds; Lieutenant-Colonel Henry, Captains Durrah and Donald Mackenzie, Lieutenants the Hon. F. Morris, H. Pigou, D. Moody, D. Rankin, and Sir William Crosby were wounded. Hostilities were soon afterwards terminated: Napoleon Bonaparte abdicated the throne of France; and in September the second battalion embarked from Ostend for Scotland, and landed at Leith in November.

The war in Europe having terminated, the *first battalion of the Royal North British Fusiliers* was selected to proceed to America, in consequence of Great Britain having become involved in war with the United States; it embarked from Genoa on the 12th of May, and arrived at *Gibraltar* on the 7th of June; and on the 11th, sailed with the Twenty-ninth and Sixty-second Regiments for the West Indies, where it joined the corps under Major-General Robert Ross. The fleet, with the troops on board, sailed from *Bermuda* on the 3d of August, and proceeded to the *Bay of Chesapeake*, when the American flotilla fled for refuge up the *Patuxent River*. To ensure the capture or destruction of this flotilla, the troops landed at the village of *St Benedict*, from whence they advanced to the delightful village of *Upper Marlborough*, when the Americans destroyed their flotilla to prevent its falling into the hands of the British. The object of the expedition had thus been accomplished, but the army had advanced within sixteen miles of *Washington*, and the enemy's force was ascertained to be such as would authorise an attempt to carry the capital. The troops, accordingly, advanced on the 23d of August, routed some detachments on the road, and, encountering the American army under General Winder at the village of *Bladensburg*, gained a decisive victory over a force more than twice their own numbers, and occupying a position deliberately chosen. The light company of the regiment distinguished itself on this occasion; it had two men killed, Captain Robert Rennie, Lieutenant James Gracie, and eleven rank and file wounded.

Advancing from the field of battle, the regiment moved towards *Washington, and was the first corps which entered that city;* it was fired upon by the Americans, and had sixty-eight men killed and wounded; but all resistance was soon overcome: the arsenal, docks, and other public property were set on fire, and the conflagration of burning buildings illuminated the sky during the night, while the exploding magazines shook the city, and threw down houses in their vicinity. Having completed this service, the British troops marched back to St Benedict, and re-embarked on board the fleet.

Early on the morning of the 12th of September the troops landed at *North Point*, and advanced towards *Baltimore*, a division of Americans having fled from an entrenched position which they were preparing across a neck of land. Continuing to advance, the troops entered a closely wooded country, where they encountered a party of Americans; and Major-General Robert Ross, mixing among the skirmishers, was mortally wounded; the command of the army devolved on Colonel Brooke.

Six thousand Americans, with six pieces of artillery, and a corps of cavalry, were discovered in a position in *Godly Wood*. The light brigade extended and drove in the American skirmishers; the Forty-fourth, a party of marines, and a body of seamen from the fleet, formed line behind the light infantry; the TWENTY-FIRST, commanded by Major Whitaker (Colonel Paterson commanding a brigade), and the second battalion of the marines, formed column in reserve; and the Fourth regiment made a flank movement to turn the enemy's left. The signal was given, the British troops rushed to the attack, and in fifteen minutes the American army was driven from the field with severe loss.

The regiment had Lieutenant Gracie and fifteen rank and file killed; Major Robert Kenny, Lieutenant John Leavock, two sergeants, and seventy-seven rank and file wounded.

Colonel Paterson was commended in the public despatch for the steady manner in which he brought the brigade into action.

At two o'clock on the following morning the march was resumed, and in the evening the troops arrived at the foot of the range of hills in front of *Baltimore*, where 15,000 Americans occupied a chain of palisaded redoubts, connected by breastworks, and defended by a numerous artillery. Trusting to the innate valour of his little army, which did not amount to one-third of the numbers of the enemy, Colonel Brooke made preparations for storming the hills after dark; but, having received intimation from the fleet that the entrance of the harbour was closed up by vessels sunk for that purpose, and that a naval co-operation against the town and camp was impracticable, the enterprise was abandoned. The troops retired three miles on the following day, and then halted to see if the Americans would venture to descend from the hills; but, though so superior in numbers, they had no disposition to quit their works, and the British returned on board the fleet.

The season for active operations having passed, the fleet quitted the American coast, and the Twenty-first proceeded to *Jamaica*, where they were joined by a strong detachment (from the second battalion), commanded by Major Alexander James Ross.

An attempt on *New Orleans* was afterwards resolved upon. The fleet again put to sea, and on the 10th of December anchored off the coast of *Louisiana*, opposite the *Chandeleur Islands*, from whence the troops were removed in boats to *Pine Island*, in *Lake Borgne*, where they were stationed, exposed to a heavy rain by day and frosts by night, until the 22d of December, when the division proceeded in open boats to a desert spot about eight miles from New Orleans, where the regiments landed, and marched to a field on the banks of the *Mississippi*. The TWENTY-FIRST followed, and arrived in time to take part in repulsing a night attack by a very superior force of Americans, when the regiment had Captain William Conran and two rank and file killed; one sergeant, two drummers, and eight rank and file wounded, two men missing.

The army afterwards moved forward, and encountered many local difficulties. The Americans assembled a numerous force in extensive fortified lines and batteries, with armed vessels on the river; the advance was checked, and some loss sustained. The FUSILIERS had Lieutenant John Leavock wounded; also several men killed and wounded.

1815. Arrangements were made for attacking the enemy's fortified lines at *New Orleans* on the 8th of January, and the FUSILIERS were appointed to take part in this service. Several circumstances occurred to delay the attack, which was made under numerous disadvantages. The troops, however, rushed forward with great gallantry, and a detachment of the Fourth, Twenty-first, and Ninety-fifth (now the Rifle Brigade), captured a battery; but they were exposed to a dreadful fire, which brought them down by hundreds. Major-General Sir Edward Pakenham was killed; Major-General Gibbs and Keane were dangerously wounded; and success being found impracticable, the surviving officers and men withdrew from the unequal contest. Many officers and soldiers who had been foremost in the attack, and penetrated into the enemy's works, were made prisoners.

Major J. A. Whitaker, Captain Robert Rennie (Lieut.-Colonel), Lieutenant Donald M'Donald, two sergeants, and sixty-five rank and file of the *Twenty-first* were killed; Colonel William Paterson, Major Alexander James Ross, Lieutenants John Waters and Alexander Geddes, six sergeants, and 144 rank and file wounded; Lieutenant James Brady, Ralph Carr, and Peter Quin wounded and taken prisoners; Major James M'Haffie, Captain Archibald Kidd, Lieutenants James Stewart, Alexander Armstrong, John Leavock, and J. S. M. Fonblanque, eight sergeants, two drummers, and 217 rank and file prisoners; total loss, 451 officers and soldiers.

The capture of New Orleans appearing to be impracticable, the troops returned on board the fleet. *Fort Bowyer* was afterwards captured, but hostilities were terminated by a treaty of peace, and the regiment returned to the *West Indies*, from whence Major Pringle sailed for England, on leave of absence, and the command devolved upon Major Quin.

After a short stay at *Bermuda*, the regiment sailed for England; it arrived at Portsmouth in May, and afterwards sailed to Cork, where it landed in June.

In the spring of this year Bonaparte had returned to France and gained temporary possession of that kingdom ; but his numerous veteran legions were overpowered by British valour at *Waterloo* on the 18th of June. The British army had, however, sustained severe loss, and the first battalion was selected to proceed to the *Continent.* It embarked from Monkstown on the 5th of July, landed at *Ostend* on the 17th, and proceeding up the country under Lieutenant-Colonel Maxwell, joined the army commanded by Field-Marshal the Duke of Wellington, at Paris.

1816. Having been appointed to remain on the Continent, and to form part of the Army of Occupation in *France*, the regiment marched to *Compiègne*, and occupied several villages in the neighbourhood of that place.

On the 13th of January the *second battalion was disbanded at Stirling*, transferring the men fit for duty to the first battalion.

Towards the end of January the regiment was removed to *Valenciennes*, and in October was reviewed, with the Army of Occupation, by Field-Marshal the Duke of Wellington.

On the death of General the Hon. William Gordon of Fyvie, Lieutenant-General James Lord Forbes was appointed Colonel of the regiment, from the Fifty-fourth Foot, by commission dated the 1st of June.

1817. A considerable reduction being made in the British contingent of the Army of Occupation, the regiment proceeded to *Calais*, where it embarked for England, and landed at Harwich on the 2d of April.

1818. In May the regiment marched to Portsmouth. In June the officers were authorised to wear a long coat of a pattern approved of by His Royal Highness the Duke of York.

1819. The regiment embarked at Portsmouth under the command of Lieutenant-Colonel North, C.B., in March, for the *West Indies*, when Major-General Lord Howard Effingham, then commanding the district, in a letter to the colonel commanding, expressed the high opinion he had of the regiment, with reference to discipline, conduct, and appearance, while serving under his command.

The regiment landed at *Barbadoes* in April.

1820. In September a detachment of 100 rank and file proceeded to *Tobago*, where it remained until January 1821, during which period it lost four officers and thirty-seven men by an epidemic disease.

1821-1822. The regiment left Barbadoes in March 1821, when seven companies proceeded to *Demerara*, under the command of Major Leahy, and three to *Berbice*, under Major Champion.

In August 1821 the regiment sustained a severe loss in the death of Lieutenant-Colonel John M. North, C.B.; he was succeeded in command by Lieutenant-Colonel John Thomas Leahy.

1823. Insurrectionary movements having been made by the negroes in the district of *Mahaica, Demerara*, in August the Fusiliers, under Lieutenant-Colonel Leahy, were employed in reducing the revolted slaves to obedience, in which they succeeded. For their excellent conduct on this occasion they received the thanks of Lieutenant-General Sir Henry Ward, K.C.B., commanding the Windward and Leeward Islands; of the Court of Policy of the Colony; of His Royal Highness the Duke of York, the Commander-in-Chief; and of His Majesty King George IV.

1824-1825. From Demerara the headquarters were removed to St Vincent in January 1824, and received the thanks of Major-General Murray previous to embarking. At the same time, the Court of Policy voted, as a special and permanent mark of the high estimation in which the inhabitants of the colony held the services of Lieutenant-Colonel Leahy, the officers, and soldiers, "FIVE HUNDRED GUINEAS TO BE LAID OUT IN THE PURCHASE OF PLATE FOR THE REGIMENTAL MESS," and "TWO HUNDRED GUINEAS FOR THE PURCHASE OF A SWORD FOR LIEUTENANT-COLONEL LEAHY;" also, "FIFTY GUINEAS FOR THE PURCHASE OF A SWORD FOR LIEUTENANT BRADY," who commanded a detachment at Mahaica, and whose cool, steady, and intrepid conduct, aided by the courage and discipline of his men, gave a speedy and effectual check to the progress of the revolt in that quarter.

The gift of 500 guineas was expended in providing a *silver centrepiece, which, after sixty-one years, still adorns the mess table of the first battalion of the regiment.*

1826-1827. In December 1826 and January 1827, the regiment embarked from St Vincent and Grenada for England, after a service of eight years in the West Indies, during which period it had lost by disease fourteen officers and 400 men. Previous to quitting those islands, it received an expression of approbation and thanks from Admiral Sir Charles Brisbane, G.C.B., governor of St Vincent; from the Council of that island, and from the Commander of the Forces in the Windward and Leeward Islands. It landed at Cowes, in the Isle of Wight, in January, February, and March 1827, and was removed to Windsor, where it had the honour of doing duty at that place during His Majesty's residence.

1828. From Windsor the regiment was removed in the spring to Winchester, and afterwards to Portsmouth; it was subsequently stationed at Bath, and in October embarked from Bristol for Ireland; it landed at Waterford, from whence it proceeded to Fermoy.

1829-1830. The regiment was removed in June 1829 to Mullingar, and in May 1830 the headquarters proceeded to Kilkenny, with detachments at Carlow, Athy, Maryborough, and Wexford.

1831. In September the regiment marched for Dublin, where it embarked for England in October, and landing at Liverpool, proceeded to Weedon.

1832. It was removed to Chatham.

1833. During the years 1832 and 1833, the regiment embarked by detachments in charge of convicts for New South Wales, Australia, and Van Dieman's Land (now the island of Tasmania), the last detachment arriving at Hobart Town on the 2d of December 1833.

1834-1838. During the years from 1834 to 1838, the Fusiliers were employed throughout the island of Tasmania, and at Perth, Port Phillip, Swan River, and Western Australia, on detachment duty in charge of various convict stations, and parties on public works; only two companies, with band and staff, remaining at headquarters. The duties were incessant, hard, and very trying, but, on all occasions, performed in such a manner as to meet the approbation of the Government.

1839-1841. The headquarters embarked from Hobart Town in February 1839, arriving at Calcutta in May, and proceeded to *Chinsurah*, where it remained until May 1840. In August, headquarters and five companies proceeded by boats up the river Ganges to *Dinapore*, under the command of Major Picton Beete, followed shortly afterwards by the remaining four companies, under the command of Lieutenant-Colonel George Deare, who had joined from leave of absence.

On the concentration of the regiment at Dinapore, Colonel Deare assumed command, and for the first time for eight years the whole corps met together, much to the delight of those who had been so long separated.

The regiment remained at Dinapore until November 1842. During its stay at this place it suffered severely from cholera.

1842. Early in November, the regiment marched for *Agra*; on arrival at *Buxar*, its destination was changed for *Kamptee*, in the *Madras Presidency*, where it arrived on the 6th February 1843.

1843-1844. The regiment was stationed at Kamptee.

1845-1846. Disturbances having occurred in *Punjaub*, and a large army of Sikhs having assembled to attack the British, the Fusiliers, after a station of nearly three years at Kamptee, were ordered to proceed with all speed to the *North-West Provinces*, to strengthen the British force.

They commenced their long and trying march on the 6th of December, with sanguine hopes of arriving in time to take part in the active operations of the field.

On arrival at *Agra* on the 7th February 1846,—*after having marched the previous thirty-four days without the interval of a halt,*—to the great disappointment of all ranks, the regiment was ordered to occupy barracks at that station.

This step, it appears, was found necessary on account of the Rajah of *Gwalior* having shown some symptoms of disaffection towards the British Government. It was, therefore, considered unadvisable to remove it further from Gwalior, being the only European regiment in the North-West Provinces not actually in the field.

At this particular time no regiment could have been in a more efficient state—a body of some eleven hundred men, thoroughly acclimatised, of splendid physique, in robust health, and inured to hard work by camp life, with long and constant marches—*they were fit to go anywhere, and do anything.* Some idea may be formed of the regiment's fitness to take the field from the circumstance that only fourteen men were in hospital, most of whom would in case of emergency have been able to bear arms.

1847. Leaving Agra on the 15th of January, the regiment arrived at *Cawnpore* on the 1st of February.

Orders having been received for return to England, and, as usual with all corps leaving India, its ranks were opened for volunteering to other regiments. With the temptation of a liberal bounty, 396 of its best men were induced to accept the offer ; and on the 1st of November the Fusiliers commenced their homeward march of 622 miles to *Calcutta,* completing the journey on the 30th of December.

On arrival at Calcutta, the offer for volunteering was renewed, and a further number of eighty-six accepted it, making a total loss to the regiment of 482 non-commissioned officers and men—*a body second to none, men of the finest physique, first-class characters, good soldiers, and thoroughly disciplined*—.

While the regiment lay encamped on the glacis of *Fort-William,* awaiting embarkation, it was called upon to take part in the reception ceremonial of the Earl of Dalhousie, the newly appointed Governor-General of India.

1848-1849. In January 1848, the embarkation of the regiment commenced; headquarters, with three companies, proceeding on the 28th by the ship "Monarch," which, under the temporary command of the second lieutenant-colonel, J. T. Hill, arrived at Gravesend on the 11th May; the remainder of the corps, in three detachments, followed in due course by other ships—the last, the "Tudor," arriving on the 3d of June.

It proceeded to Canterbury as its first home station, after having been absent on foreign service for sixteen years.

Lieutenant-Colonel Deare having proceeded home by private ship, now joined and resumed command.

Consequent upon the large number lost by volunteering to remain in India, recruits had been raised to fill the vacancy to the number of 470, who were awaiting the arrival of the regiment at Canterbury.

In July sudden orders were received for the removal of the Royal North British Fusiliers to Edinburgh, to which place they proceeded by railway, and were quartered in Edinburgh Castle, furnishing detachments for Berwick-on-Tweed, Greenlaw Military Prison (now called Glencorse), and Leith Fort.

Shortly after its arrival the regiment sustained a severe loss by the retirement of Lieutenant-Colonel George Deare, who had commanded it for the previous ten years.

Any historical record of the Twenty-first Royal Scots Fusiliers would be very incomplete in which there was no mention of Colonel George Deare, or no reference to his work in the regiment.

This officer obtained his first commission into the Fusiliers, and he remained in it until his retirement from the army in 1848. He was its colonel and commanding-officer during the last ten years of his service.

When he succeeded to the command in 1840, after his return from leave in England, he found the state of the regiment, in regard to drill and general discipline, very unsatisfactory. The cause of this was, that for the previous eight years the greater part of the regiment was broken up into small detachments, and scattered far and wide over Tasmania and Australia ; several large drafts of recruits joined soon after the landing of the regiment in India ; the men of these drafts had no knowledge whatever of battalion drill, nor had they yet acquired the spirit and tone of soldiers.

This was the material that Colonel Deare had to begin his work of reorganisation upon. He set about his labour almost single-handed ; the only one who could render him such aid as he thought worth accepting was Captain, afterwards Colonel, Ainslie, who was mortally wounded at the battle of Inkerman.

He began at the beginning—first squad, and then company drill. He was a perfect master of drill, even to its most minute details ; he went from squad to squad, or from company to company, directing, correcting, and instructing officers, non-commissioned officers, and men. By the end of the drill season of 1840 the regiment was fairly well drilled, and by the end of the succeeding season it could not be surpassed, perhaps not equalled, by any regiment in the service.

If he did not entirely originate, he resuscitated, developed, and matured an admirable *Regimental System* which succeeding commanding-officers have been careful to maintain.

He was the *beau ideal* of a "commanding-officer ;" about six feet in height, well proportioned, and of such a dignified and soldierly bearing as to inspire respect. His word of command was surprisingly loud, clear, and "*ringing ;*" he had an eye like an eagle, that could detect any unsteadiness or inattention in the ranks, front or rear, from No. 1 to No. 10 company ; and, as he knew the name of almost every man in the regiment, the offenders were at once indicated by name. It ought also to be mentioned that he was an excellent horseman, and was always well mounted.

As an "*orderly-room colonel*" he was unequalled; possessed of clear perception, and an unerring judgment, he was quick to discern the false from the true in any matter brought before him; and no judge that ever sat on the bench was more free from bias or prejudice in his decisions. He was what is called a "*light punisher*" of the crimes that soldiers are ordinarily guilty of, but anything mean, unmanly, or unsoldier-like, was visited by him with scathing reproof and well-merited punishment.

The men were proud of "*George*" (their pet name for him), and they *knew* that he was proud of them; they believed that their own regiment was the best in the army, and they were not mistaken.

It is not, therefore, to be wondered at that such an *esprit-de-corps* raised and maintained the character of the ROYAL NORTH BRITISH FUSILIERS to such a height, that even the ruinous *short-service system* has not been able wholly to destroy.

1850. After having been stationed nearly two years in Edinburgh, the regiment received orders to move to Glasgow.

During the regiment's unusually long stay in the capital of Scotland, it maintained the highest character for its conduct, discipline, and *esprit-de-corps;* securing for all ranks, the greatest respect and kindness from the various social grades of the civil community, and cementing more strongly the connection of the corps with the land of its creation, by introducing into its ranks a goodly number of Scotch recruits.

Colonel Deare's retirement gave the command to Major Peddie, an officer of long regimental connection, *members of his family having served successively in the Royal North British Fusiliers from its first formation.*

Colonel Peddie, shortly after his promotion, exchanged with Colonel T. G. Brown from the Forty-first regiment.

On the 5th of April the regiment proceeded to Glasgow, under the command of Lieutenant-Colonel T. G. Brown; two companies, under Major F. G. Ainslie, were detached to Paisley, and one subaltern, one sergeant, and twenty rank and file to Dumbarton Castle.

1851. The regiment remained in Glasgow until April, when it received orders to move; the headquarters to Newcastle-upon-Tyne, two companies to Carlisle, two to Sunderland, and two to Tynemouth Castle.

While the regiment was in Glasgow, an officer, Lieutenant Clemison, was accidentally killed by a fall from his horse.

Before leaving for England, the Major-General commanding the troops in North Britain, conveyed to the regiment an expression of his approbation on account of its high state of efficiency and general good conduct whilst in Scotland.

1852. The regiment moved in February, the headquarters to Hull, one company to Scarborough, three to Bradford, and three to Leeds.

On the 1st of March the establishment was augmented to 850 rank and file.

On the 2d of April a detachment proceeded from Leeds to Barnsley.

1853. On the 5th of January a company proceeded from Leeds to Sheffield, to occupy new quarters.

By the retirement of Lieutenant-Colonel Thorpe, the command of the regiment devolved upon Major Ainslie, who succeeded to the vacant lieutenant-colonelcy.

During the station of the regiment at Hull, an experimental issue of 100 rifles was made; the pattern was that known as the *Minie*, carrying a conical bullet.

Colonel Ainslie, to encourage *esprit-de-corps*, instituted a system of granting distinctive badges of various descriptions to be worn by the best shots in the regiment. These, for the ten best shots in each company, consisted of *cross muskets*, worn on the upper right sleeve of the fatigue jacket; the best shot in each company had his surmounted by a *grenade*, and the best shot in the regiment had the *grenade surmounted by a crown.* No money prize accompanied these badges, *the honour of wearing them was considered sufficient.*

It will thus be seen that the "*Schools of Musketry*" *subsequently established were anticipated by the* "*Fusiliers.*"

1853. In June the regiment moved to Dublin, *via* Liverpool; headquarters to Ship Street Barracks, four companies to Aldborough House, and two to Beggars' Bush Barracks.

An Industrial Exhibition having been opened at Dublin, Her Majesty and the Prince Consort visited that city in August. On this occasion she reviewed the whole of the troops in garrison.

In October the regiment was inspected by Major-General Cochrane, commanding the Dublin District, and received much praise from him for its steadiness, good conduct, and smartness.

The head colonel of the regiment, General the Right Hon. Sir Frederick Adam, G.C.B., G.C.M.G., died on the 24th of August, and was succeeded by Major-General Sir De Lacey Evans, K.C.B.

1854. By Horse Guards letter, dated 7th January, Her Majesty was pleased to authorise the word "*Bladensburg*" to be borne on the colours and appointments of the ROYAL NORTH BRITISH FUSILIERS. In March orders were received to hold the regiment in readiness for such active service as might be required.

On the 10th of May orders were given for an augmentation to twelve companies, with a total strength of 1400 non-commissioned officers and men.

In June positive orders were issued to prepare for immediate embarkation for Turkey.

On the 3d of August the regiment proceeded to Cork for embarkation for the seat of war in Turkey, leaving its depôt, consisting of four companies, at Buttevant.

On its departure from Dublin, in addition to receiving expressions of high approval as to conduct, appearance, and discipline, from the military authorities, the press of that city was unanimous in its comments upon its uniform good conduct during its stay, and its splendid appearance.

On arrival at Cork, it was inspected by Major-General

Mansfield, commanding that district (an officer not easily pleased).

He inspected most minutely Nos. 1 and 2 and the front rank of No. 3 Companies, when he ceased, exclaiming aloud,—"*That will do, Colonel Ainslie, close your ranks, and march past; I never inspected such a regiment!!*"

With regard to the Fusiliers on this occasion, the Cork *Times*' correspondent writes, "*The equal of the Fusiliers never paraded in Cork barrack square, not even excepting the Guards!!*"

On the 15th of August the regiment embarked on board the "Golden Fleece," sailed for the Black Sea, joined the army in *Varna Bay* on the 4th of September, and formed part of the Fourth Division, under the command of Lieutenant-General Sir George Cathcart.

The Allied Army sailed on the following morning, and reached *Old Fort* on the 14th of September.

The regiment bivouacked on the seashore, where it landed. A night of heavy rain followed, and being without any shelter, both the clothes worn by the men and the change carried in their folded greatcoats were thoroughly drenched,—an ominous introduction to the subsequent hardships of the campaign.

During the few days of its stay here, cholera broke out, and many sick had to be re-embarked.

On the 19th of September the army commenced its march in the direction of *Sebastopol*. During the advance this day the regiment suffered severely from heat, want of water, and sickness; many men fell out of the ranks in consequence. The army bivouacked on the banks of the *Bulganac*.

At daylight, on the morning of the 20th of September, the division of which the Fusiliers formed a part was again under arms, advanced in support, and at noon came in full view of the Russian army strongly posted on the *Heights of Alma*.

The action commenced about two o'clock, and was fought in about two hours and a half. The regiment bivouacked on the heights for the night. The two following days (21st and 22d) were occupied in removing the wounded, and burying the dead.

On the morning of the 23d, the regiment resumed its march in the direction of the river *Katchka*, on the banks of which it arrived about three P.M., and bivouacked for the night. On the 24th, the march was continued towards the *Belbec River*, on the banks of which a halt for the night was made.

On the 25th the Fourth Division and the 4th Light Dragoons remained on the Belbec to protect the communication with the Katchka, and to allow the sick being sent to that place for embarkation on board the fleet.

The remainder of the Allied Army, in the meantime, moved forward. On the 26th, the march was resumed in the direction of "*Mackenzie's Farm*," passing through a very thick wood ; the farm was reached about three o'clock, and after descending to the banks of the "*Tchernaya*," the troops bivouacked for the night.

The following day (the 27th) the regiment marched to the *heights above Sebastopol*, its permanent camping-ground during the whole siege.

On the morning of the 25th of October a staff officer galloped suddenly into camp, and conveyed an order to the commanding-officer directing the regiment to proceed with all despatch to the plain of *Balaklava*, to act in conjunction with the other regiments of the Fourth Division, to repel an attempted attack by the Russians upon that place. No time was lost, each regiment proceeded as quickly as possible ; the distance was about six miles ; and nearly the whole of it was performed at the "*double.*" The regiment remained in the field the whole of the day, but the battle was one of cavalry and artillery, the only infantry engaged being the NINETY-THIRD SUTHERLAND HIGHLANDERS.

The Fourth Division returned to its camp in the evening.

On the 5th of November the ever memorable battle of *Inkerman* was fought—"*The Soldiers' Battle*," as it has been termed, where a Russian army, numbering as ten to one compared with the British, was completely defeated with great loss.

To give full details of the engagement and movements would in an historical record occupy too much space; suffice it to say, the FUSILIERS on this occasion greatly distinguished themselves.

The right wing was commanded by Lieutenant-Colonel F. G. Ainslie, until mortally wounded, when Lieutenant-Colonel F. P. Haines succeeded; the left wing was under Lieutenant-Colonel Lord West.

The gallantry displayed by every man of the regiment was beyond all praise. Kinglake, in his description of the battle of Inkerman, speaks of them as "*those superb Fusiliers*" and "*most magnificent troops.*" During the action, when broken up into small parties, they held the most forward and important positions, and mere handsful of them charged and repulsed large battalions of Russians.

The regiment suffered severely in this action. Of 402 men who went into the field, seven officers, six sergeants, and 114 men were killed or wounded; Lieutenant E. Hurt, killed; Colonel Ainslie, mortally wounded; Captain G. N. Boldero, Lieutenants Stephens and Killeen, severely; Lieutenant King, dangerously; and Lieutenant Templeman, slightly.

Many incidents might be recorded of individual gallantry, as, for instance, that of Private Patrick M'Guire, who, being considerably in advance of his comrades after making a charge, and finding himself isolated, took shelter behind a rock, where he was shortly attacked by three Russians, who attempted to take him prisoner. After a desperate struggle, M'Guire succeeded in shooting one, bayonetting another, and brought in the third as his prisoner. For this spirited action, he was awarded the medal for gallant conduct in the field. Another noteworthy incident. Lieutenant Hurt, who fell in advance, and was supposed for a time to be dead, was observed by Colonel Haines to be still alive. He called the attention of Colour-Sergeants Rutherford and Higdon, who were then firing from under cover, to the fact, and asked them to move out and bring in the wounded officer. A most withering fire of grape-shot rained over this part of the field at the time, yet these brave sergeants, regardless of all danger to themselves, at once ran out and lifted their wounded officer, but their effort was of no avail, as he was struck by another shot and killed in their arms.

It may not be out of place to record here a few words regarding Lieutenant Roger Killeen, who was promoted to a commission after a lengthened service in the ranks. In all positions, he was respected as a man of high aspirations, an indefatigable and conscientious soldier, and with a soul of honour. Although severely wounded while carrying the regimental colour, he, from devotion to his regiment, and a feeling of determination, declined to leave the field, and still held on until the action was over. He continued to do his duty throughout the siege, and never gave way, but, by his example, encouraged his men to perform their duties with stubborn pertinacity. Noble old man! He retired with the rank of major, and is since dead.

The regiment sustained a severe loss by the death of Lieutenant-Colonel Ainslie (the last but one of the old Fusiliers). Although a strict commanding-officer, he was a brave man, ever exhibiting the greatest anxiety to maintain the character of his regiment; his untimely death was much felt, more especially when the services of men of his stamp were so much needed.

The command now devolved upon Major John Ramsay Stuart, the only survivor of *the old Fusiliers*, he having always served in the Twenty-first.

On the morning of the battle of Inkerman a great number of the non-commissioned officers and men had only been relieved from duty in the trenches, where they had been for the previous twenty-four hours; they were just in time to take part in the action.

The total strength of the enemy in the field, actually engaged, was 71,841, with 271 guns; while the British force only numbered 7464, afterwards reinforced at the end of the action by 8219 French.

The operations being all over, the regiment returned to camp at five P.M.

The following tribute to the character of Lieutenant-Colonel Lord West, appears in Kinglake's *Invasion of the Crimea*, and holding as he did, a distinguished position on this occasion, the extract will, no doubt, be found interesting:—"Lord West (the late Earl De la Warre) commanded a wing of the Twenty-first Fusiliers, the regiment to which Colonel Haines belonged. In anticipation of a probable conflict on Mount Inkerman, these two gifted officers had the forethought to go over the field some days before, and it may be inferred that the important part they both took in the action was in some measure owing to a knowledge of the ground thus wisely acquired beforehand. Although communicating to me full information on other objects, the late Earl abstained from volunteering any statements of the part he had taken in bringing the *Inkerman Battle* to its final crisis, and I am assured this abnegation of self, conjoined with fearless assumption of power already recorded, was thoroughly characteristic of his exalted nature. 'A splendid soldier,' writes one brother officer of him, 'no truer gentleman, no more honest or braver man ever lived.'"

14th November 1854.—On this day the army suffered from an enemy of another kind,—a storm, the equal of which has rarely been experienced. The tents almost all were blown down, all had to take shelter from the pelting storm of wind, rain, and sleet, behind low stone ditches, so low, as only to give shelter to men in a crouching position. Fortunately, the wounded of Inkerman had been sent to Scutari five days previously. This day began the winter's miseries and sufferings of the Crimean Army, so much spoken of, and ever to be remembered by those who experienced them.

After this, hardship and troubles every day increased; clothing began to wear out without any prospect of renewal; rations became scarce, in consequence of the difficulty experienced in bringing them from Balaklava. The roads between it and the camp, a distance of seven miles, had become so heavy and bad as to be nearly impassable; it was not unusual for the mules bringing up the provisions to sink in the mud, taxing the combined strength of the drivers to extricate them. At last, even the mules failed, a sufficient number could not be had to perform this duty, and in the absence of them, the regiment had to detail fatigue parties to take their place.

In many cases, after spending a miserable night in the trenches, up to their knees in sludge or half frozen to death, these men had to go to Balaklava for food, which frequently consisted of a scanty ration of biscuit and rum.

With scarceness of food, clothing, and fuel, sickness soon did its fatal work amongst the regiment. The ration of salt meat had to be used in its raw state, the allowance of green coffee had now become useless, there being no means to roast or grind it, consequently it was left on the ground unused after being issued to the men.

Some idea may be formed of the intensity of the cold, from the fact that the men's moustaches during the night became solid lumps of ice, and their whiskers froze to their threadbare blankets.

Although the privations suffered by the regiment were very great, they were not so great as that of some other corps; and this was mainly, or entirely, owing to the indefatigable exertions made by the Regimental Quartermaster's Department.

From early in October 1854 up to the end of the siege, the regiment was busily occupied, either as working parties constructing batteries and approaches, or guarding the same; when not so employed it furnished fatigue parties for bringing rations from Balaklava, or carrying shot and shell to the trenches. So continuous and excessive was the labour, that the men in an average were *eleven out of fourteen* nights in the trenches, either as working or covering parties. The misery and monotony of trench duty was sometimes enlivened by sorties by the Russians.

On the 22d of March 1855, when the regiment furnished working and covering parties, a sortie was made in force all along the line, and after some severe fighting the enemy were driven back at every point, leaving many dead and wounded; the loss of the regiment amounted to four killed and seven wounded.

On the 4th of April 1855, when under a heavy cannonade, a shell fell into the trench; Private William Watt, No. 5 Company, with great presence of mind, took it up and threw it over the parapet, when it exploded. The act was performed in the coolest manner, and did not seem to the actor as anything worth talking about.

On the night of the 7th June 1855, strong parties from the British and French advanced to the attack on the *Mamelon* and *Quarries;* the Twenty-first Fusiliers furnished a strong party under Lieutenants J. G. Image and S. H. Clarke. The attack continued the whole night without intermission, resulting in the capture of the quarries, with a loss to the regiment of one killed and two wounded.

After the capture of the quarries it was decided by a council of war, that the works should be stormed and the town taken, the 18th of June being fixed for the attack.

On the evening of the 17th of June, the Fusiliers paraded on *Cathcart's Hill* to be "told off" for their different duties on the morrow; one hundred men were detailed to carry woolsacks to fill up the trench in front of the great Russian Redan, an equal force was allotted for carrying and planting escalading ladders, the remainder of the regiment formed a storming party. The men were then dismissed to take what rest they could, having been ordered to parade again at midnight.

Exactly at twelve o'clock P.M., the parade was again formed, each man furnished with one hundred rounds of ammunition. All were in readiness to march off, when Lord West, the commanding-officer, addressed his men as follows:—"Fusiliers, you will have hard work to perform, but I have every confidence in you. I know that you will retain the honour which you gained on the field of Inkerman, and whoever he may be who is spared to come out of this night's work, if we meet in after-years, will not be lost sight of by me. In conclusion, have your hearts in your hands ready to dash them over the parapet and follow them." Each party then marched off to take the position in the *advanced trench*, to await the signal for attack, which was given just as day was breaking; the woolsack and ladder parties cleared the parapet at a bound, never stopping until they reached the Russian trench; this was accomplished under a deadly fire from the enemy's guns, which had opened all along the Russian line at a given signal.

Close behind came the storming parties led by General Sir John Campbell, up to the mouths of the enemy's guns, only to be shot dead beneath them. At this point, through some mismanagement, the reinforcements did not come up, consequently the attack failed; after holding their ground in the enemy's trench for upwards of half-an-hour, the parties had to retire under a most terrific fire of grape and canister. The regiment suffered very severely on this occasion.

On the night of the 15th August 1855, a sortie in force upon the advanced trenches was frustrated by the vigilant alertness of a party of the regiment under the command of Lieutenant R. C. Winsloe; the object of the intended sortie, doubtless, was to prevent assistance being rendered to the French and Sardinians on the Tchernaya, on whom an attack had been made.

On the death of Lord Raglan, General Simpson, who succeeded to the command, ordered another attack to be made on the Russian position on the 8th September 1855. The Fusiliers, on this occasion, were in reserve. The attack was made simultaneously at twelve noon; the French attacking the *Malakoff*, and the British the *Great Redan*,—the former driving out the Russians, who retired into the Redan, and strengthened the place still further against the British attack.

The fight continued till late in the afternoon, when the British troops were ordered to retire, the Fourth Division and Highland Brigade being directed to take up position in the British advanced trenches, ready to renew the attack on the following morning; but, as the Russians had evacuated the Redan during the night, the services of these troops were not required.

With the fall of the Malakoff and Redan, and the retirement of the Russians across the harbour to the north side, Sebastopol was captured, and its siege concluded.

On the 7th of October 1855, the Fusiliers and four other regiments of the Fourth Division, under the command of Brigadier the Hon. A. Spencer, and an equal force of the French army, were embarked on board the Allied Fleet in Kamiesh Bay, under sealed orders; the Fusiliers, on board H.M.S. "Hannibal," flagship of Admiral Sir Houston Stewart, K.C.B., and commanded by the Right Hon. Captain C. D. Hay, C.B.

The expedition sailed the same night, and two days later arrived in front of *Odessa*, taking up a position before the arsenal.

This was only a feint, and intended to draw off the troops from Kinburn, the intended point of attack. The feint succeeded. On the morning of the 14th the Allied Fleets steamed off from Odessa towards *Kinburn*, which was reached on the 15th. The troops were immediately landed upon a desolate, sandy spit, and encamped in line in front of the fortifications. The fleets opened fire, and after a bombardment of about two hours, the whole of the works surrendered; 1420 prisoners were taken, and eighty-one pieces of heavy cannon; forty-five killed and 130 wounded Russians were also found in the forts. *With the bombardment of Kinburn the war terminated.*

On the 12th of November, the regiment returned to its camp before Sebastopol, where it remained until after the *declaration of peace.*

The Fusiliers, during their stay of one year and eight months in the Crimea, lost by killed, wounded, sickness, and missing, 623 of all ranks.

The regiment, under the command of Lieutenant-Colonel

J. R. Stuart, embarked at Balaklava for Malta, where it arrived on the 2d June, and occupied quarters in that island until the 17th March 1860.

1860. On the 17th of March the regiment embarked on board the troop-ship "Himalaya" for conveyance to the West Indies. It arrived at the island of Barbadoes, and disembarked on the 6th of April, relieving the Forty-ninth regiment, Two companies, under the command of Major H. Gray, were detached to *Demerara*. The detachment at Demerara had not been long there when a terrible fire broke out, destroying about one-fourth of *George Town*. The services of the Fusiliers on this occasion were of great value in saving life and property, for which they received the thanks of the Field-Marshal Commanding-in-Chief, also those of the Governor and General in command of the Windward and Leeward Islands.

In June 1860, Demerara was visited by *yellow fever*. The detachment of the regiment was removed into camp at Belfield, where it remained until 1861.

1862. In September headquarters at *Barbadoes* was also visited by *yellow fever*, and in a day or two it committed sad havoc. But for the decided measures caused to be taken by Surgeon A. J. Greer, who insisted upon having the regiment moved into camp, the loss by deaths would probably have been very great.

To carry the removal into effect was no easy matter, owing to the Commissariat being entirely deficient of the means of transport. This department divested itself of all labour and responsibility by transferring its duty to the regiment.

This duty had, of course, to be carried out by the Quartermaster, Mr Grahame (now Major Grahame), who himself was slowly recovering from an attack of the fever. This officer received from the Commissariat full authority to employ for this purpose all the available transport in the island.

The labour of collecting this transport devolved upon the Quartermaster himself. In two hours after receiving this authority, the removal began, and in less than two days the whole regiment, with tents, beds, bedding, camp equipage, regimental and personal baggage, had been removed to *Gun Hill,* a distance of seven miles.

After the removal of the regiment, no further case of fever occurred. Thus that it escaped this dreadful scourge is entirely owing to the promptitude and decision of Surgeon A. J. Greer, and the energy of Quartermaster George Grahame.

To show the virulence of the disease, and the great necessity there was for the removal, it may be stated that the officers of the West India regiment, thinking the Fusilier officers' quarters were more healthy, removed into them, and in the two following days six of them fell victims to the awful epidemic.

The men, while in camp, were not allowed to remain in their tents in a state of idleness; and to usefully employ them, Colonel Stuart obtained permission to construct a winding road up the face of the hill leading to the camp, wide enough for two carts to pass each other, thereby saving a lengthened journey around the hill. This road, when completed, was appropriately named Fusilier Road.

On 11th October No. 7 Company, under the command of Captain Bruce, proceeded to the island of *St Vincent* to repress an insurrection of the native population, and returned to headquarters on the 3d May 1863. For this duty it received the thanks of the Governor and General in command of the Windward and Leeward Islands.

1863. Another very destructive fire occurred at George Town, Demerara, and the detachment of the Fusiliers was called upon to render assistance. By the advice of the officer in command, and with the approval of the local authorities, it was considered necessary to blow up a certain building in order to arrest the progress of the flames. Lieutenant Hutton, a promising young officer of the regiment, volunteered to perform the hazardous duty, and, in carrying it out, lost his life by the explosion.

1864. The regiment embarked for England on the 5th of August, on board H.M.S. "Tamar," having been relieved by the second battalion Third regiment, "The Buffs."

Prior to embarkation, his Excellency Sir James Walker, Governor of the Windward and Leeward Islands, and Major-General Brooke, commanding the troops, both conveyed to the regiment expressions of their high sense of the discipline and conduct of the Fusiliers during the four years of their stay in the command.

The regiment disembarked at Portsmouth on the 15th of August, and occupied quarters in Anglesea barracks. A few days after its landing, it was inspected by Major-General Lord William Paulet, commanding the district, who expressed himself highly pleased with its appearance; and added, "Your good name has preceded you, and you are more like a regiment that has been on home service for ten years, than one just returned from foreign service."

The following month it was again inspected by his Royal Highness the Duke of Cambridge, Field-Marshal Commanding-in-Chief, in the barrack square; and, on the same afternoon, as a portion of the Portsmouth division, at a general parade. On both of these occasions his Royal Highness spoke of the regiment in the highest terms of praise.

1865. On the 10th of April, the regiment removed from Portsmouth to North Camp, Aldershot. During its short stay of eight months at this station, it maintained its usual high character for discipline and appearance; so much so, that, an emergency having occurred for the services of a regiment at Glasgow, to relieve the Sixty-third, suddenly ordered to Ireland, to the surprise of all, the Twenty-first Fusiliers were selected, although it had been the *last* to join the camp. This feeling of surprise was so great, that in reference to the telegram ordering its removal, the "camp authorities" referred back the question, "*Was it not the Thirty-first, not the Twenty-first?*" The telegraphic reply came back in the form of a question, "*When was the Thirty-first made Fusiliers?*"

The regiment accordingly proceeded by train to Glasgow on the 28th of December, and occupied the Gallowgate barracks, furnishing detachments to Paisley and Ayr.

1866. While stationed at Glasgow, the *Fenian movement,* in Ireland, became somewhat serious. The Fusiliers were ordered to that country on the 10th of September, to assist in restoring order; and, on arrival at Dublin, were located in Richmond barracks.

1867. March 1, on a parade in Phœnix Park, Dublin, at which every available man of the regiment was present, Colonel John Ramsay Stuart, C.B., bade farewell to the corps, in which he had served for upwards of thirty-seven years, the last thirteen of which as its colonel and commanding-officer.

He was the last of the officers of the ancient *régime.* During his career he had endeared himself to all ranks by his blunt soldierly, natural manner, and the performance of many kind and considerate acts.

As a commanding-officer, during most trying times, his experience, knowledge of human character, and tact in dealing with men, earned for him—*and most justly, too*—the reputation of a wise and prudent officer, to whom all under his command looked up with confidence for advice, support, and assistance.

His devotion to the interests, character, and *esprit de corps* of the home of his youth and mature age was intense, and he always succeeded, whether in camp, garrison, or quarters, in gaining for the regiment the high position of being *second to none* as regards conduct, appearance, and discipline.

The portly figure, genial countenance, and stentorian voice of "Old Donald," or "Ramsay," as he was indiscriminately but affectionately called, were, for a long time after his retirement, much missed on parades, "hauls up," and inspections. All these things considered, no old soldier need be told that the parting between colonel and men was very affecting.

His memory will ever be kept green by the "old hands"

of the Fusiliers, as one who was as a father to the regiment. By many subsequent acts of kindness and substantial help to old Fusiliers, he has shown that his affection for the corps remains unabated.

The command of the regiment now devolved upon Lieutenant-Colonel J. T. Dalyell.

In this year also the regiment suffered another loss, in the transfer of Surgeon A. J. Greer to the Seventeenth Lancers.

This officer served with the regiment for the previous fifteen years, as assistant-surgeon and surgeon. His skill, care, and sympathy for the sick will never be forgotten by those who have been under his treatment. A patient felt that he had in Dr Greer not only a doctor, but a friend. This was notably manifested in the trying circumstances of the Crimean campaign.

Medical officers in those days knew their men, and the men knew them.

The Royal North British Fusiliers, under the command of Lieutenant-Colonel J. T. Dalyell, left Dublin for the Curragh Camp, on the 9th of July.

On the 15th of October, the regiment was removed to Enniskillen, furnishing detachments to Newry, Drogheda, and Armagh.

1868. On the 13th June it was again quartered in the Curragh Camp, and employed on election duties from the 17th to the 27th of November. When orders were received for the regiment to move from Enniskillen to the Curragh Camp, the inhabitants, on account of its universal good conduct, and the kind and cordial intercourse that had existed between them and all ranks, during the period of its stay, memorialised the military authorities at Dublin, to permit the Fusiliers to remain amongst them for another year.

Singularly enough, a somewhat similar action was taken by the inhabitants of this town, when the regiment was stationed there sixty-seven years before (*vide record for the year* 1801).

1869. On the 15th of February the regiment embarked at Queenstown, on board H.M. Indian troopship,

"Serapis," for the East Indies; arriving at *Alexandria* on the 26th, it disembarked and proceeded by railway across the desert to *Suez*, and again embarked on board H.M. Indian troopship "Jumna," arriving at Bombay on the 20th of March; on the 22d, the regiment was transhipped into three smaller vessels for conveyance to *Kurrachee*, where it arrived on the 26th, and remained stationed, furnishing a detachment of two companies to *Hyderabad*.

During the nine months' station of the regiment at Kurrachee, it suffered very severely from fever; from this disease the deaths amounted to sixteen men, eleven women, and forty-one children.

Fever prevailed to such an extent, and its effects were so disastrous, that the regiment was frequently unable to furnish sufficient healthy men for its own guards. Under these circumstances, it was found necessary to remove the Fusiliers to a more healthy station; *Bangalore*, in the *Mysore Territory*, was selected, to which place it removed on the 25th January 1870.

1870. The removal to Bangalore was of the greatest benefit to the regiment; in the course of a few months the health of all ranks had materially improved.

The change from the arid sandy plains of Kurrachee was marvellous; and by the end of the year, the Fusiliers had attained their usual energy.

A regimental theatre, weekly lectures and concerts, workshops, coffee-shop, bakery, and aerated-water manufactory were established. A splendid sward and a mild temperature made cricket in the mornings and evenings an agreeable and healthy pastime to the lovers of that game. Rifle matches were much in vogue, and during the three annual Bangalore rifle meetings of 1871–1873, the grand trophy of each year fell to the good shooting of the Fusiliers, in addition to many other prizes in the various competitions, amounting to a considerable value.

A regimental newspaper, *The Dekhnewalla*, was also established, and ably conducted by Mr G. Holland, the regimental schoolmaster.

From time immemorial the regiment had, more or less, Highland pipers, although unauthorised; but for the previous twenty years, from one cause or another, it had been without them; in this year, however, means were adopted for training some men, and in a short time the regiment had ten fully equipped pipers playing at its head.

On the 22d of September No. 2 Company, under the command of Captain R. Cook, proceeded to *Cannanore* on detachment duty, returning to headquarters on the 25th of November.

1871. On the 16th of December No. 3 and 7 Companies, under the command of Captain Edgell, proceeded on detachment to *Trichinopoly*, returning to headquarters on the 23d February 1872.

1872. On the 2d of December the regiment proceeded to Fort St George, Madras, to relieve its second battalion, which was under orders to return to England. [This was the first occasion on which both battalions met.]

Two companies proceeded to Trichinopoly in relief of a detachment belonging to the second battalion.

1873. On the 1st of April the new *localisation scheme* was published, by which the "ROYAL NORTH BRITISH FUSILIERS" became the "AYRSHIRE COUNTY REGIMENT," with the depôts of both battalions stationed in Ayr barracks, under the title of the *Sixty-first Brigade Depôt or Centre*, since changed to the *Twenty-first Regimental District.*

1874. The regiment formed part of a camp of exercise, which took place at Bangalore; at its conclusion, the Commander-in-Chief of the Madras Presidency, Sir F. Haines, complimented the Fusiliers on their high state of efficiency.

1875. On the 26th of February the Royal North British Fusiliers, under the command of Lieutenant-Colonel J. T. Dalyell, embarked at Madras for *Rangoon*, where they arrived on the 5th of March, and relieved the Forty-fifth regiment (Sherwood Foresters).

On the 27th of March a detachment, under the command of Captain C. Patterson, proceeded, *via* the river

Irrawady, to *Bassein;* Captain Patterson died while upon this duty.

On the 12th of December a detachment of 130 men, under the command of Captains T. E. Stuart and W. N. Carey, proceeded to *Port-Blair* in the *Andaman Islands*, to take charge of convicts.

1876. By general order, dated 22d of December, an establishment of one pipe-major and three pipers was officially authorised for each battalion of the Royal North British Fusiliers.

1877. On the 14th of December the regiment moved from Rangoon to Secunderabad, where it arrived on the 27th of the same month.

1878-1880. In January 1878, Colonel J. T. Dalyell having obtained the command of a brigade depôt in Scotland, the command of the first battalion of the Fusiliers devolved upon Lieutenant-Colonel A. Templeman.

The loss of Colonel Dalyell was much regretted by all ranks.

During a service of thirty-one years in the *old corps*, which he had joined as a second lieutenant, he ever maintained the character of a good officer, and a kind, unassuming gentleman.

The eleven years the regiment had the good fortune to be commanded by him will long be remembered in its annals as most pleasant and happy ones.

He endeared himself to all by his quiet, firm, just, and dignified manner. An excellent drill, good disciplinarian, and kind friend, he always took the most active interest in everything that concerned his regiment. His care for the comfort of the wives and families of his non-commissioned officers and men was unceasing.

Colonel Dalyell was the last commanding-officer of the Fusiliers who held that position under the old rules, which fixed no limit to the tenure of command. He has since attained the rank of major-general on the active list.

1881. During this year, by general orders of 11th of April and 1st of May, the title of the regiment was altered from Royal North British Fusiliers to that of THE ROYAL SCOTS FUSILIERS.

The uniform was also changed from the ordinary line tunic to a "Highland doublet," with dark "tartan trews;" the officers' lace to be of the "thistle pattern," and the ordinary sword to be replaced by the "claymore."

The regiment embarked for England on the 4th of November; landed at Portsmouth on the 1st December, and proceeded to Dover.

During the thirteen years the battalion served in India it lost by death four officers, one bandmaster, one colour-sergeant, nine sergeants, eight corporals, and 137 privates.

1882. A Horse Guards letter was received, dated 29th of April, authorising scrolls, bearing the names of the following engagements, to be borne on the regimental colours, viz.:—"*Blenheim*," "*Ramillies*," "*Oudenarde*," "*Malplaquet*," and on the 20th of September another letter authorised "*Dettingen*" in addition to the foregoing.

Colonel A. Templeman, having completed the period of limited regimental command, retired upon half-pay, and was succeeded by Lieutenant-Colonel G. F. Gildea, from the second battalion.

In August Major Grahame retired from the regiment, and from the service. He was the oldest Fusilier in the regiment, and the oldest quartermaster, but one, in the army.

The retirement of this officer demands very much more than a passing notice in what may be called a popular regimental record.

Major Grahame is a born Fusilier. His father, Mr William Grahame (now of Hobart in Tasmania), entered the regiment in 1817. He is still alive and well, and, being between eighty and ninety years of age, is probably the oldest living Fusilier. It is also probable that he is the only survivor of those of the regiment who took part in the suppression of the insurrection of the negroes in Demerara in 1825. Although separated from his old corps for nearly half a century, he takes an intense interest in all that concerns it. Such is the stuff the old Fusiliers were made of. From this it will appear that father and son form by themselves a continuous, unbroken, historic chain, stretching over nearly seventy years of the regiment's history.

Major Grahame served thirty-seven years in the Twenty-first; the last twenty-seven as its quartermaster.

In a former part of this record mention has been made of the indefatigable exertions of the quartermaster's department to keep the regiment supplied with food and clothing in the Crimea, but it will bear to be briefly referred to again.

It may be confidently and truthfully asserted the comparative efficiency of the Fusiliers for "trench duty," and any immunity from the most dreadful sufferings, were mainly due to the efforts of the quartermaster's department. The labour of this officer was an unceasing daily struggle to keep the gaunt wolf "hunger" from eating out the vitals of the regiment. With him a difficulty was a something to be overcome. But he could not work miracles to procure that which was unprocurable ; and when pinching and relentless hunger gnawed within empty stomachs, there was at least the satisfaction of knowing that all had been done by the quartermaster that effort and forethought could accomplish.

This daily struggle to keep life in the regiment was carried on unobtrusively, and although of the utmost importance, there was nothing in it to catch the eye, and inspire the pen of "our war correspondent ;" or the *glorious*, to furnish material for an eulogistic paragraph in an official despatch. It brought no "decorations" nor emoluments, but it *deserved them.*

At Major Grahame's suggestion, a "Regimental Library and Recreation Room" were established, also a "School Committee," that a stimulus might be given to self-culture, education, and sobriety. This was done years before the War Office, or Horse Guards, introduced such institutions.

The quartermasters, riding-masters, and sub-inspectors of schools of the army, owe a debt of gratitude to Major

Grahame for his persevering, laborious, and successful efforts in obtaining for them an improvement in their status, by which they were granted higher rank and increased pay while serving, with a somewhat more liberal retiring allowance.

By his brother officers in the regiment Major Grahame was greatly esteemed. A gentleman in principle, in manners, and education, he at all times reflected credit on the position he held. His constant aim was to add by word and deed to the comfort and well-being of the regiment, and in doing so, he earned for himself a name (which will not soon be forgotten) worthy of honour from all old Fusiliers.

Major Grahame served under seven different regimental commanding-officers, in various parts of the world—viz., two East Indian tours; one Mediterranean and West Indian; in Tasmania; and throughout the whole of the Crimean campaign of 1854–5–6, including the battles of Alma, Inkerman, siege and fall of Sebastopol, also capture of Kinburn. Medal with three clasps, and Turkish war medal.

1883. On the 19th of November the battalion, under the command of Lieutenant-Colonel Gildea, left Dover for Aldershot.

1884. By the retirement of Lieutenant-Colonel Gildea, the command of the first battalion of the Royal Scots Fusiliers devolved upon Lieutenant-Colonel E. T. Bainbridge, who at present commands (June 1885).

1885. During this year the regiment moved from Aldershot to Portland, where it at present remains (June 1885).

Part 2.

THE SECOND BATTALION.

Part 2.

THE SECOND BATTALION.

Historical Record of the Second Battalion of the Twenty-first Royal North British Fusiliers, now Second Battalion Royal Scots Fusiliers.

1857. The Government having considered it necessary to increase the strength of the British army, orders were consequently issued for the formation of SECOND BATTALIONS, which were to be added to all regiments up to, and including, the Twenty-fifth. Thus the Royal North British Fusiliers had a second battalion added, the battalions being designated first and second of the Royal North British Fusiliers, now known as the *Royal Scots Fusiliers.*

1858. The formation of the second battalion of the Fusiliers took place in the month of April, at Paisley, under the superintendence of Colonel E. Last, from the Ninety-ninth regiment, this officer having been appointed as its first commanding-officer.

A number of officers from the first battalion joined on promotion to a higher grade; also, a party of non-commissioned officers and men transferred their services from the first battalion to aid in its formation; and being all old soldiers, recently returned from the Crimea, formed a nucleus, which introduced the excellent system of regimental interior economy so long in use by the parent battalion.

On the 19th of December, the second battalion of the Royal North British Fusiliers proceeded to Newport, in Wales.

1859. The regiment was removed to Aldershot on the 12th of August.

In this year Colonel Last retired from the command of the battalion, and was succeeded by Lieutenant-Colonel Lowe, from the Thirty-second Light Infantry.

1860. In October, the headquarters and right wing proceeded to Shorncliffe, the left wing to Dover.

1861. In May the left wing rejoined headquarters from Shorncliffe.

1862. On the 2d of April, the battalion left Shorncliffe for Portsmouth, for embarkation for Dublin; on arrival it was quartered in Beggars' Bush barracks.

Colonel Lowe exchanged with Colonel Robertson, from the first battalion of the Sixth, who assumed command.

Colonel Lowe's departure from the Royal North British Fusiliers was much regretted, he having, during his three years of command, gained the respect of all under him.

1863. In June the battalion, under the temporary command of Lieutenant-Colonel G. N. Boldero, proceeded from Dublin to the Curragh Camp. Almost immediately after the arrival of the battalion at this station, where it remained only one month, orders were received to hold itself in readiness for foreign service.

On the 19th of July it embarked at Kingstown for the East Indies; and, after a voyage of four months, landed at Madras, and was stationed as follows:—Headquarters, *St Thomas' Mount*, with detachments at *Fort St George, Madras*, and *Arcot.*

1864-1865. The battalion occupied the above quarters for about a month, when it was ordered to *Bellary*, where it arrived on the 13th January. While at this station a severe cholera epidemic occurred, which carried off a considerable number of men, women, and children.

1866-1867. On the 1st of January 1866 the second battalion Royal North British Fusiliers marched for *Secunderabad*, where it arrived on the 1st of February.

1868-1869. In October 1868 the battalion marched from Secunderabad to Madras, where it embarked, under the temporary command of Major A. Templeman, for *Burmah.*

Colonel Robertson, having been appointed to the staff of the Indian Army as Adjutant-General at Madras, the command of the battalion devolved upon Lieutenant-Colonel E. A. T. Steward.

1870-1871. By the retirement of Lieutenant-Colonel Steward, on account of ill-health, Lieutenant-Colonel F. Lyster, from the first battalion, succeeded to the command.

The battalion, during its stay in *Burmah,* was stationed two years in *Rangoon,* furnishing a detachment of 120 men for convict guards at *Port-Blair,* in the *Andaman Islands.* In the third year of its stay it was located by wings at *Thyetmo* and *Tonghoo ;* the *right* being at the former, and the *left* at the latter station.

In December 1871 the battalion, under the command of Lieutenant-Colonel F. Lyster, returned to Rangoon.

1872. The battalion embarked for Madras in January, and on arrival occupied barracks in Fort St George, detaching two companies to Trichinopoly to relieve the detachment of the first battalion, the headquarters of which were stationed at *Bangalore.*

On the 1st May a very destructive cyclone occurred at Madras, causing much loss of life and property amongst the shipping in the harbour. A number of vessels were forcibly driven from their anchors, and stranded upon the beach, exposed to terrific wind, with a heavy rolling surf.

The Fusiliers turned out, and rendered all the assistance in their power in rescuing the crews, passengers, and securing property. They put forth the most strenuous and successful efforts ; and their arduous and dangerous exertions called forth the highest terms of approbation from the Government of Madras, the general commanding, as well as from the merchants and inhabitants. These latter, as a permanent memorial of the regiment's services on this occasion, presented it with a massive silver vase of splendid appearance, and excellent workmanship, as an ornament for the officers' mess table.

The battalion left Fort St George on the 3d of December for embarkation for England, *via* Bombay, having been relieved by its first battalion from Bangalore.

1873. In January the battalion landed at Portsmouth, and at once proceeded to Scotland, and was stationed—headquarters at Stirling Castle, three companies at Perth, two companies at Dundee, and two companies at Hamilton.

1874. The Fusiliers proceeded to Glasgow in May. In this year the battalion, under the temporary command of Lieutenant-Colonel G. F. Gildea, proceeded to Aldershot.

Consequent upon the retirement of Lieutenant-Colonel F. Lyster, Lieutenant Colonel W. P. Collingwood, from the first battalion, succeeded to the command.

1875. In July the battalion was removed to the Portsmouth District and occupied Portsdown Forts.

1876. The second battalion Royal North British Fusiliers removed into Portsmouth in July, and occupied quarters in Clarence barracks.

1877. In November the battalion returned to Scotland, and was stationed in Fort-George, detaching two companies to Dundee.

1878. In April the second battalion of the Fusiliers was removed to Ireland—headquarters being stationed in Richmond barracks, furnishing a detachment of two companies to Ship Street barracks.

On the 28th of September the battalion was removed from Dublin to the Curragh Camp, and shortly afterwards received orders to hold itself in readiness for embarkation.

During the short time which intervened before embarkation the battalion was augmented to its full strength for foreign service by the addition of a large draft of well-trained soldiers, under the command of Major Hazlerigg, from the Sixty-first brigade depôt at Ayr.

1879. On the 20th of February the second battalion Royal North British Fusiliers left the Curragh for Cork, under the command of Lieutenant-Colonel W. P. Collingwood, and embarked at Queenstown for South Africa.

As the ship steamed into *Simon's Bay*, she struck upon the *Roman Rocks*. Such an incident was well calculated to cause alarm and confusion on board. The good discipline of the Fusiliers was shown on this occasion by the steadiness and ready obedience to orders, which called forth the praise of their commanding-officer, Lieutenant-Colonel Collingwood, and Captain Fulton, commanding the ship.

Consequent upon this accident, the battalion was transhipped to another vessel, and proceeded to *Durban*, where it disembarked on the 31st March.

In anticipation of active operations, the band exchanged their instruments for rifles, the only musicians left being the pipers.

On the 3d of April the Fusiliers left Durban for "up country," arriving at *Pietermaritzburg* on the 5th, where two companies, under the command of Captain F. Willoughby, were left for the defence of the town.

Resuming their march on the 8th of April, they arrived at *Dundee* on the 23d.

On the 2d of May they left Dundee, and on the 30th joined the division under the command of Major-General Newdigate at *Koppie-allein*.

On the 1st of June the division crossed *the Blood River into Zululand*.

On the 3d the battalion resumed its march to *Ity-oty-Ozi River*, close to the place where the Prince Imperial of France was killed.

On the 4th, on arrival in camp, the division constructed a fort, which was named "Fort Newdigate," in honour of the general.

During the night an attempt to surprise the position was made by the Zulus; but the enemy being discovered, after a few rounds, they disappeared.

Here (Fort Newdigate) two companies of the Fusiliers, with two guns, and one troop of dragoons, were left as a garrison, the remainder of the battalion proceeding on the 6th to the *Upoko River*, where, after shelling the bush, the division formed " laager."

On the morning of the 7th of June the first brigade, consisting of the second battalion of the Royal North British Fusiliers, Fifty-eighth regiment, with cavalry and artillery, went out and cleared the bush, destroying the native "kraals," and carried off a large quantity of "mealies." Here the battalion remained until the 16th, awaiting the arrival of General Wood's column from *Conference Hill* with stores. After this party had joined, the march was resumed along the banks of the Upoko.

On the 17th the Fusiliers were again employed constructing another fort, afterwards named Fort Marshall; two companies of the battalion with two guns, and a squadron of the Seventeenth Lancers, being detailed to remain behind to garrison it, under the command of Lieutenant-Colonel Collingwood.

On the 18th the remainder of the battalion, commanded by Major Hazlerigg, resumed its march towards *Ulundi*, arriving on the 30th of June.

On the 4th of July, early in the morning, the troops crossed the *Umvolosi* river, and continued their march towards Ulundi.

The enemy were not visible until the scouts had ascended the opposite heights; immediately afterwards, they were seen swarming upon the hills all around.

The division advanced in "square," the Seventeenth Lancers in rear of the rear face; the enemy descended from the hills and surrounded the square, making their fiercest attack upon the rear face where the Fusiliers were posted, with the evident intention of getting between the troops and their camp, and so cutting off retreat.

The enemy was ultimately repulsed; the number of their dead found lying in front of the position defended by the Fusiliers, bore evidence to the coolness and accuracy of aim of the latter.

In this engagement none of the Fusiliers were killed, but Major Winsloe and a number of men were wounded.

On the 5th July, Lieutenant-Colonel Collingwood having completed the regulated period of five years in command of the battalion, was succeeded by Lieutenant-Colonel Gildea, who, being in England, the command was temporarily held by Major Hazlerigg.

Lieutenant-Colonel Collingwood, in giving up the command, expressed in orders the great regret he had in severing his connection with the regiment in which he had served for a period of twenty-nine years.

On the 26th of August the battalion marched for *Pretoria.*

On the 7th of September, the two companies under Captain Willoughby, which had been left at Pietermaritzburg, rejoined headquarters.

On the 8th of September the following general order was published:—" The General commanding desires it to be notified to all wounded in hospital, that the Queen has most graciously commanded, through Lady Frere, her Majesty's particular inquiries respecting them."

The battalion halted at *Wakkerstroom* until the 15th of September, owing to the deficiency of baggage animals; when the right half battalion under Major Hazlerigg, marched for *Heidelberg*, the remaining half battalion under Major Bainbridge, being unable to proceed for want of transport. In a few days mules were procured, and it resumed the march.

On the 25th of September, when at Standerton, orders were received for the formation of a body of *mounted infantry*, to consist of two officers, two sergeants, two corporals, and forty-six privates. The officers selected were Lieutenants Collings and Lindsell, and twenty-five non-commissioned officers and men from each half battalion.

These men did excellent service throughout the whole campaign, not only when engaged with the enemy, but also as scouts and on outpost duty, performing the duties of cavalry in a most efficient manner.

October 15th.—Having arrived at Heidelberg, the battalion received orders to proceed to *Middleburgh*, at which place it arrived on the 20th, and was there joined by the Ninety-fourth regiment; remaining at Middleburgh until the 24th, when it marched for *Fort Webber*, arriving there on the 20th of November.

Near this place "the Transvaal Field Force" was formed. The battalion marched from Fort Webber, and arrived at *Fort Albert Edward* on the 23d of November.

On the following day two companies of the Fusiliers were attached to form part of an advanced post, to procure water for the troops.

On the 26th the remainder of the battalion, with the column, marched from Fort Albert Edward. This was a very distressing day's march owing to unused roads, made nearly impassable by recent heavy rains; great delay was caused by the waggons sticking fast in the soft muddy soil; during this night the troops endured great discomfort from a tremendous thunderstorm, accompanied by torrents of drenching rain.

On the 27th a live ox was given the battalion; it was soon slaughtered, cut up, and distributed. Whilst each man was cooking his portion as best he could, the "fall in" sounded. Even the hardships the men were suffering did not prevent the scene from assuming a ludicrous aspect; men rushing into the ranks with their half-cooked ration on the point of their bayonets! This was the only food they had during the day. After marching from an early hour until seven P.M., the troops encamped on the banks of a river opposite *Sekukuni Town*, about two miles distant.

November 28th.—At two A.M. tents were struck, the battalion paraded, and the whole column advanced across the river to the attack.

In the storming, capture, and destruction of *Sekukuni strongholds*, the Fusiliers took a prominent part; the casualties being three men killed; Captains Willoughby and Gordon, and sixteen men wounded.

December 3d.—Orders were published by General Wolseley and Colonel Russell, in which the troops were complimented on their steadiness, endurance, and gallantry.

December 14th.—In consequence of a Boer rebellion, the Fusiliers received orders to proceed to *Pretoria*, where they arrived on the 22d.

Two harassing and toilsome campaigns, during which the battalion marched upwards of one thousand miles, had reduced the men's clothing to a very dilapidated condition. Each man endeavoured as best he could to repair the rents and holes in his apparel; but the material obtainable for patching was most unsuitable, being neither the colour nor texture of the garment itself. Patches of biscuit-bags, blankets, waterproof sheets, may cover deficiencies, but do not add to the splendour of a soldier's uniform! Whilst all were much alike, the general appearance of the Fusiliers did not cause much remark amongst themselves; but when they marched into Pretoria, the worn-out condition of the clothing, with its many-coloured patches, contrasted grotesquely and ludicrously with the neat, clean uniform of the soldiers in garrison.

Although the clothing was worn-out, the wearers were not, the men being healthy, and soldierlike in their bearing.

The battalion did not long continue in this healthy condition; the reaction, the excitement of active warfare, together with bad water, brought on enteric fever, and in a short time 127 men were in hospital with this disease.

1880. On the 17th of January, the following extract from the *London Gazette* was published in orders:—

"The Queen has been graciously pleased to approve of the following promotions being conferred upon the under-mentioned officers, in recognition of their services during the late Zulu campaign; dated 29th November 1879:—Brevet, to be Lieutenant-Colonels—Major Arthur Grey Hazlerigg, 21st Foot; Major Richard William Charles Winsloe, 21st Foot."

On the 24th of January, the following order was published:—

"The Queen commands me to express her satisfaction at the news of taking *Sekukuni Town*, as well as her sorrow for the loss of brave officers and men, and her anxiety for the well-doing of the wounded. The General Commanding is well assured that the troops in South Africa will appreciate this mark of Her Majesty's gracious sympathy with her soldiers."

On the 3d February, whilst in camp at Pretoria, a tremendous hurricane and thunderstorm, accompanied by torrents of rain, occurred; tents and huts were blown down, and the men exposed to the full force and inclemency of the elements. Some men were injured, but not seriously.

May 27th.—Lieutenant-Colonel Gildea, who had joined from England, and assumed command of the battalion, was appointed commandant of the garrison.

On the 16th of July, the following order was published by Major Bainbridge, in temporary command of the battalion, consequent upon the death of Brevet Lieutenant-Colonel A. G. Hazlerigg from fever:—

"It is with deep regret that the commanding-officer has to announce to the battalion the death of Brevet Lieutenant-Colonel Hazlerigg, which occurred this morning at Pretoria barracks. He feels sure his loss will be greatly felt by all ranks of the Fusiliers, with whom he served so many years, always having had their welfare and interest at heart; and that, therefore, his memory as a kind and gallant officer will long be retained amongst all ranks of the Royal North British Fusiliers, a regiment he was so fond and proud of."

September 16th.—It was intimated to the troops that the Queen had been graciously pleased to command that medals be granted for services in South Africa.

November 14th.—A force, consisting of one division of guns (N, 3 R.A.), commanded by Lieutenant Randall; one company of the Royal Scots Fusiliers, under Lieutenant Falls; and a half troop of "mounted infantry," under Lieutenant Lindsell,—the whole under the command of Major Thornhill, Royal Artillery,—marched from *Pretoria*, to be stationed at *Potchefstroom*, where a "fort" was built. This force was afterwards augmented by D Company,

Royal Scots Fusiliers, under Lieutenant Browne, from *Rustenberg*, leaving the latter garrisoned by E Company, under the command of Captain Auchinleck.

In December the *Transvaal War* broke out, and at the commencement of it the regiment was distributed as follows:—*Pretoria*—Headquarters, with A, B, F, and H Companies, and a half troop of mounted infantry, under the command of Lieutenant-Colonel Gildea. *Potchefstroom*—C and D Companies, with half troop of mounted infantry, under the command of Lieutenant-Colonel Winsloe. *Rustenberg*—E Company, under the command of Captain Auchinleck.

Two drafts from England were detained from joining while hostilities were in progress, one having been ordered to occupy *Pietermaritzburg*, under the command of Major Bainbridge, and the other at *Schoon-Hoogte*, under the command of Captain Whitton.

The first active operations commenced at Potchefstroom on the 16th of December, and at Pretoria on the 19th of December.

On the 12th of December, Lieutenant-Colonel Winsloe joined the troops at Potchefstroom, and assumed command of the whole force, amounting to 213 of all ranks, with two nine-pounder guns of the Royal Artillery, occupying at the time the "fort," "gaol," and Landroost's office, with the two guns in shallow "gun-pits;" whilst the Boers held the town.

On this day (the 16th of December) the first shots were interchanged between the British and the Boers, in a slight skirmish; the mounted infantry and Boers having come into collision on outpost duties. One of the enemy was wounded. Later the attack became general, more especially upon the Landroost's office, which was situated in the market square. The attack did not long continue, the enemy retiring with some loss.

The strength of the enemy at this time was about 800 mounted men, subsequently augmented to 1400, all excellent shots, and armed generally with the Westley-Richards rifle.

The night of the 16th was spent in strengthening the position; the numbers of the garrison was increased, but not strengthened, by the accession of twenty-one women and children, who sought protection within the British lines.

On the 18th, the detached portion of the force in the Landroost's office was compelled to surrender, the position being untenable. During its occupation, Captain Falls, one of the most promising officers of the battalion, was killed, and several of the men wounded. A "flag of truce" was arranged for carrying out the evacuation; but, while it was still flying, the enemy opened fire, wounding one man.

It was also found necessary to abandon the "gaol," and the party occupying it retired, after dark, carrying their wounded on stretchers made with rifles. During this operation one man was killed.

The fort, which was only twenty-five yards square, now sheltered the whole British force, including women, children, horses, and mules. It was now invested on three of its sides.

19th.—The garrison had no water within the fort, save what was obtained from a well, which had been sunk to a depth of thirty feet; but this only yielded nine gallons a day.

For the first few days, whilst the fort was not closely invested, water was obtained outside, at a distance of about 1200 yards; but, as some men and horses were either killed or wounded in these watering operations, this mode of relief had to be discontinued. Fortunately, on this day a heavy rainstorm gave sufficient water to last until the 21st; and on this latter day another rainfall renewed the water-supply for three days longer. On the 21st, to lessen the demand for food and water, the whole of the horses and mules were turned adrift.

A second well was sunk, which fortunately yielded a plentiful supply, equal to the wants of the garrison throughout the siege.

1881. *January 1st.*—The enemy opened a heavy fire from their Westley-Richards, and, in addition, brought into play a ship's gun. The latter was soon silenced by the fire of the nine-pounders in the fort. This lasted about two hours.

Several men were disabled, whose services could ill be spared from such a small force.

The defences of the fort consisted chiefly of sand bags, which were made by the sick and wounded from the tents and other materials available during the operations. The whole of the tents, except five reserved for the sick, were appropriated to this purpose—men, women, and children, all living and sleeping in the "open."

January 22d.—Long before the fort had been invested on all sides, a trench, which had been opened some 220 yards distant, became a source of annoyance and loss. A sortie of an unusual character, as much as it took place in open daylight, was determined upon. A party of an officer (Lieutenant Dalrymple Hay), with a sergeant and ten men, made a rush across the open ground for the trench, but, before reaching it, three of the number fell; the remaining seven dashed at the trench, which held eighteen of the enemy. Four were taken prisoners, eleven fell either killed or wounded, and three made their escape.

Next day, under a flag of truce, the enemy sent a doctor to attend their wounded. The garrison courteously lent stretchers to convey them into town. On the following day these stretchers were returned, laden with fruit and some medicinal appliances for the wounded. Such acts of kindness help to relieve warfare of some of its horrors.

February 4th.—Long ere this the troops had been put on short allowance of food. This allowance had to be gradually decreased, and at length there was nothing left but "meallies" (Indian corn) and "Caffre corn" (millet). The supply of these was totally inadequate, especially for men who had to work hard day and night. The imperfect manner of preparation and cooking brought on diarrhœa, dysentery, and enteric fever.

February 10th.—This morning the enemy opened a heavy fire on all sides of the fort, which was continued during the whole of this and the following day. During these two days sixty-five shots from the ship gun struck the fort. Considering the heavy fire, the casualties were very few.

March 19th.—The food being exhausted, and the sick and wounded dying for want of nourishment, it was found impossible to hold out any longer. Accordingly, a flag of truce was sent to the Boer commander, asking for an interview, and on the 21st of March terms of capitulation were signed—the garrison to march out of the fort with all the honours of war, flags flying, and drums beating.

On the 23d March the garrison marched from Potchefstroom for Natal, where it arrived on the 2d of May.

The following are the names of the officers present with the detachment of the Royal Scots Fusiliers :—

Brevet Lieutenant-Colonel R. C. WINSLOE, *commanding.*
Captain L. FALLS, *killed.*
Lieutenant P. W. BROWNE.
Lieutenant C. F. LINDSELL.
Lieutenant H. E. LEAN.
Lieutenant DALRYMPLE HAY.

The following district order was issued by Colonel Bellairs, C.B. :—

"The fort at Potchefstroom capitulated on the 21st of March, but only when its garrison was reduced to extremity, and after as brave a defence as any in military annals, the troops marching out with the honours of war, and proceeding through the *Orange Free State* to *Natal.*

"The sterling qualities for which British soldiers have been so renowned have been brilliantly shown in this instance during a long period of privation, and under very trying circumstances.

"Colonel Bellairs begs Lieutenant-Colonel Winsloe, and the officers and men under him, will accept his thanks for the proud and determined way in which they have performed their duties.

"By order.
(Sd.) "M. CHURCHILL, Captain,
Deputy Assistant Adjutant-General." *

* The foregoing account of the defence of Potchefstroom has been condensed from an article written by Lieutenant-Colonel Winsloe, A.D.C., in *Macmillan's Magazine.*

Whilst the two companies, under the command of Lieutenant-Colonel Winsloe, were besieged at Potchefstroom, five companies, with headquarters at *Pretoria*, under the command of Lieutenant-Colonel Gildea, were actively employed.

When news of the treacherous, cowardly, and sanguinary massacre of the Ninety-fourth regiment was received at Pretoria, the country was placed under martial law, and preparations for defence were made.

The garrison consisted of one division of N Battery, Fifth brigade, Royal Artillery, with two nine-pounder guns; one company of Royal Engineers, headquarters and five companies of the Royal Scots Fusiliers, with half a troop of mounted infantry; two companies of the Ninety-fourth regiment, with one troop of mounted infantry; volunteers, consisting of the Pretoria Carbineers, Nourse's Horse, and the Pretoria Rifles.

The Boers mustered in great numbers, and occupied several strong positions, but at some distance from the British lines.

Reconnaissances in strength, under the command of Lieutenant-Colonel Gildea, were made on the 19th and 28th of December 1880, 6th and 16th of January 1881, and the last in February; after which the garrison was able only to act on the defensive. Engagements, more or less serious took place on four of these occasions, the casualties being, on the British side, four killed, three officers and eleven men wounded. Amongst these was Lieutenant-Colonel Gildea, severely.

Captain Auchinleck, who commanded a company of the Fusiliers at Rustenberg in December 1880, was also dangerously wounded while defending that place.

The colours of the Ninety-fourth regiment, which had been torn off the poles to save them from capture by the Boers, on the occasion of their treacherous attack upon that corps, were conveyed to Pretoria, and handed over for safe keeping to the Fusiliers.

At the termination of hostilities, these colours were

delivered to the detachment of the Ninety-fourth, under the command of Major Campbell, then serving at Pretoria.

October 1881.—The battalion marched from Pretoria for *Durban*, where it arrived on the 3d of January 1882, embarking on the same day for the East Indies.

January 26th, 1882.—On arrival at Bombay, orders were received for the battalion to proceed to *Secunderabad* to relieve its first battalion, proceeding to England; thus completing a tour of home duty, and taking part in three campaigns in South Africa; relieved its first battalion, which, ten years before, had relieved it in the same presidency.

The second battalion, while stationed at Secunderabad, in honourable imitation of the first battalion, established a regimental newspaper, called the *Fusee*, which not only relieved the monotonous life induced by an Indian climate, but stimulated the educated portion of the battalion to amateur literary exertions.

1884. After being stationed nearly three years at Secunderabad, the battalion was removed at the latter end of this year to Burmah, where it remains at the present date—June 1885.

Appendix.

Appendix No. 1.

SUCCESSION OF COLONELS

OF

The Twenty-First Regiment of Foot,

OR

THE ROYAL NORTH BRITISH FUSILIERS,

NOW

THE ROYAL SCOTS FUSILIERS.

CHARLES, (FIFTH) EARL OF MAR.

Appointed 23d September 1678.

CHARLES, LORD ERSKINE, succeeded to the title of EARL OF MAR in 1668, on the decease of his father, John, fourth Earl of Mar; and in September 1678 he raised a regiment of foot, now the TWENTY-FIRST, or the ROYAL NORTH BRITISH FUSILIERS. He was a member of the Privy Council of Scotland, in the reign of King Charles II., and also of King James II. In 1686, he was succeeded in the command of his regiment by Colonel Buchan.

The Earl of Mar disapproved of the measures of King James II., and was about to embark for the Continent, in November 1688, when the Prince of Orange landed in England. He appeared at the Convention of the Estates assembled by the Prince of Orange; but joining the disaffected party, he was arrested. He died on the 23d of April 1689, and was succeeded in the title by his son John, sixth Earl of Mar, whose estates were forfeited in consequence of his having erected the standard of rebellion in Scotland, in 1715, in favour of the Pretender, as narrated at p. 14 of the Historical Record of the TWENTY-FIRST, ROYAL NORTH BRITISH, FUSILIERS.

THOMAS BUCHAN.

Appointed 29th July 1686.

Thomas Buchan was an officer in the Scots army, in the time of King Charles II., and rose to the rank of lieutenant-colonel in the Royal Regiment of Scots Horse, which was disbanded in 1689. King James II. promoted him to the colonelcy of the TWENTY-FIRST regiment; and he adhered to the interests of the Stuart family at the Revolution in 1688. He served in Ireland under King James, and was detached with a few men to Scotland, to support the Highland clans in their resistance to the Government of King William III. The clans were, however, not successful in their enterprises, and they submitted to the authority of King William; when Colonel Buchan retired to France.

FRANCIS FERGUS O'FARRELL.

Appointed 1st March 1689.

This officer was a decided advocate for the principles of the Revolution of 1688, and King William nominated him to the colonelcy of the SCOTS FUSILIERS, which corps he commanded in the Netherlands, under Prince Waldeck, and afterwards under the British monarch, who promoted him to the rank of brigadier-general. He served at the head of a brigade of infantry during the campaign of 1694, and was appointed governor of Deinse. He commanded the garrison of Deinse when that place was besieged, in July 1695; and was dismissed the service, by sentence of a general court-martial, for surrendering without firing a shot.

Robert Mackay.

Appointed 13th November 1695.

Robert Mackay, third son of John, Lord Reay, was an officer in the Scots Brigade in the Dutch service, and accompanied the Prince of Orange to England in 1688. He was promoted captain of the grenadier company in Major-General Hugh Mackay's regiment, and served in Scotland in 1689. He distinguished himself at the battle of Killicrankie, where he received several wounds, and was left for dead on the field of battle. He, however, showed some signs of life, and was removed to a cottage by the enemy, and eventually recovered. He was soon afterwards promoted to the rank of lieutenant-colonel, and King William gave him the colonelcy of a newly raised Scots regiment (afterwards disbanded), from which he was removed, in 1695, to the Twenty-first Fusiliers. His constitution had become debilitated by severe service and numerous wounds, and he died at Tongue, the seat of his family, in December 1696.

Archibald Row.

Appointed 1st January 1697.

This officer entered the army in the reign of King James II., and at the Revolution in 1688 he joined the Prince of Orange, who promoted him to the lieutenant-colonelcy of the Sixteenth regiment, with which corps he served in the Netherlands, and acquired the reputation of a brave and skilful officer. He served at the battles of Steenkirk and Landen, and at the siege of Namur; and was rewarded, in 1697, with the colonelcy of the Twenty-first Fusiliers. He served under the great Duke of Marlborough in 1703, and in 1704 he commanded a brigade at the battles of Schellenberg and Blenheim. On the last-mentioned occasion his brigade led the attack on the village of Blenheim, and he headed his own regiment with distinguished gallantry, advancing up to the enemy's palisades before he gave the word "fire." In a moment afterwards he was shot, and thus closed a life of honour with a death of glory. His valour has rendered his name immortal in the history of his country.

John, Viscount Mordaunt.

Appointed 25th August 1704.

John, Viscount Mordaunt, son of Charles, Earl of Peterborough, was an officer in the first regiment of Foot Guards, in which corps he rose to the rank of captain and lieutenant-colonel. He evinced great gallantry at the battle of Schellenberg, where he headed fifty grenadiers, at the storm of the enemy's works, and of that number only himself and ten grenadiers escaped. At the memorable battle of Blenheim he lost his left arm. His services were rewarded with the colonelcy of the Twenty-first Fusiliers, from which he exchanged to the Twenty-eighth regiment; but, on the death of Major-General De Lalo, who was killed at the battle of Malplaquet, in 1709, Viscount Mordaunt was reappointed to the Twenty-first regiment. He was promoted to the rank of brigadier-general on the 1st of January 1710, and died of the smallpox in April following.

Sampson de Lalo.

Appointed 26th June 1706.

Sampson de Lalo was a French gentleman of the Protestant religion, whom the Edict of Nantes forced to quit his native country. He found an asylum from persecution in England, and, entering the British army, proved an efficient and meritorious officer. After a distinguished career of service in the subordinate commissions, he was appointed lieutenant-colonel of the Twenty-eighth regiment, and was promoted to the colonelcy of the same corps in February

1704; in June 1706, he exchanged to the TWENTY-FIRST FUSILIERS. He commanded a brigade under the great Duke of Marlborough, served at several battles and sieges, and was promoted to the rank of major-general in January 1709. During the siege of the castle of Tournay, he was nominated by the Duke of Marlborough to negotiate the terms of capitulation with the governor. He evinced great gallantry at the battle of Malplaquet, where he was mortally wounded. In the *Annals of Queen Anne* it is stated that "he was in great favour and esteem in the British army."

JOHN, VISCOUNT MORDAUNT.

Reappointed 4th September 1709.

Died in 1710.

THOMAS MEREDITH.

Appointed 1st May 1710.

This officer served in the wars of King William III., who promoted him to the commission of captain in the Third Horse, now Second Dragoon Guards. On the augmentation of the army in 1702, he was nominated colonel of the Thirty-seventh regiment, then newly raised, and he accompanied that corps to Holland in 1703. In 1704 he served at the battles of Schellenberg and Blenheim, and was promoted to the rank of brigadier-general on the 25th of August 1704. In 1705 he commanded a brigade at the forcing of the French lines at Helixem and Neer-Hespen. He was advanced to the rank of major-general in 1706, and to that of lieutenant-general in 1707; in 1710 he was removed to the TWENTY-FIRST regiment, and in 1714 to the Twentieth. He died in 1719.

Charles, Earl of Orrery, K.T.

Appointed 8th December 1710.

The Earl of Orrery took an active part in raising a regiment of foot (afterwards disbanded), of which he was appointed colonel on the 1st of May 1703; in 1705 he was nominated Knight of the Thistle; and in 1706 he was removed to another regiment, afterwards disbanded. He was promoted to the rank of brigadier-general in 1709, and served at the battle of Malplaquet, at the head of a brigade of infantry, and evinced great gallantry. In 1710 he was advanced to the rank of major-general, nominated Envoy Extraordinary and Plenipotentiary to the States of Brabant and Flanders, and removed to the Twenty-first Fusiliers; in 1711 he was created a peer of Great Britain, by the title of Baron Boyle, of Marston, in Somersetshire; and in 1712 he served under the Duke of Ormond. He was sworn a member of the Privy Council in 1713. On the arrival of King George I. in England, in the autumn of 1714, the Earl of Orrery was appointed one of the Lords of the Bedchamber; he was afterwards sworn a member of the Privy Council. In 1722 he was committed a prisoner to the Tower of London, on a charge of high treason, but no crime was proved against him. He died on the 28th of August 1731.

George Macartney.

Appointed 12th July 1716.

This officer entered the army in the reign of King William III., and was promoted in April 1703 to the colonelcy of a newly raised regiment of foot (afterwards disbanded), with which he served three campaigns on the Continent, under the great Duke of Marlborough. He afterwards proceeded to Spain, and commanded a brigade of infantry at the battle of Almanza, where he distinguished himself, and was taken prisoner. In 1709 he was promoted to the rank of major-general, and in 1710 to that of lieutenant-general. His regiment having been disbanded at the Peace of Utrecht, he was appointed to the colonelcy of the ROYAL NORTH BRITISH FUSILIERS in 1716, and removed in 1727 to the Seventh Horse, now Sixth Dragoon Guards. He died in July 1730.

SIR JAMES WOOD.

Appointed 9th March 1727.

Sir James Wood served many years in the army of the States-General of the United Provinces of the Netherlands. His first commission was dated the 31st of December 1688, and he rose to the rank of brigadier-general in 1704, in which rank he was admitted into the British service, in consequence of his reputation; and in 1727 he was appointed colonel of the TWENTY-FIRST regiment. In 1735 he was promoted to the rank of major-general. His decease occurred on the 18th of May 1738.

JOHN CAMPBELL.

Appointed 1st November 1738.

John Campbell of Mamore was an officer in the army in the reign of Queen Anne, and attained the rank of lieutenant-colonel. During the rebellion in 1715 and 1716 he was aide-de-camp to the Duke of Argyll; and in June 1737 he obtained the colonelcy of the Thirty-ninth regiment, from which he was removed in the following year to the ROYAL NORTH BRITISH FUSILIERS. He commanded a brigade at the battle of Dettingen, in 1743; was appointed major-general in the following year; and during the rebellion in 1745 and 1746 he held a command in Scotland. He was advanced to the rank of lieutenant-general in 1747; removed from the Fusiliers to the Scots Greys in 1752; and in 1761 he was appointed governor of Limerick, and also succeeded to the title of Duke of Argyll. The Order of the Thistle was conferred upon his Grace in 1765. He died in 1770.

William, Earl of Panmure.

Appointed 29th April 1752.

William Maule, who had been several years an officer in the Scots Foot Guards, and a member of Parliament, was created a peer of Ireland on the 6th of April 1743, by the title of Earl of Panmure. He served at the battle of Dettingen in the same year; also at the battle of Fontenoy in 1745; and on the 1st of December 1747 was promoted to the colonelcy of the Twenty-fifth foot; from which he was removed, in 1752, to the Royal North British Fusiliers. The rank of major-general was conferred upon his lordship in 1755. In the following year he was second in command at Gibraltar; and in 1758 he was promoted to the rank of lieutenant-general. He was further advanced to the rank of general in 1770; and obtained the colonelcy of the Scots Greys in November of the same year. He died on the 4th of January 1782.

The Honourable Alexander Mackay.

Appointed 10th May 1770.

The Honourable Alexander Mackay, son of George, third Lord Reay, was appointed ensign in the Twenty-fifth regiment in 1737; and in 1745 he obtained the commission of captain in the Earl of Loudoun's newly raised regiment of Highlanders, afterwards disbanded. He served against the rebels in the same year, and was taken prisoner at the battle of Prestonpans. In 1750 he was nominated major in the Third foot, and on the 21st of December 1755 he was promoted to the lieutenant-colonelcy of the Fifty-second regiment, then newly raised, from which he exchanged, in March 1760, to the Thirty-ninth; in 1761 he was elected a member of Parliament for Sunderland; in August 1762 he was promoted to the colonelcy of the One Hundred and Twenty-second regiment, which was disbanded at the Peace of Fontainebleau; and in March 1764 he obtained the colonelcy of the Sixty-fifth. He served in America, in which country he obtained the local rank of major-general in 1768; in 1770 he received the same rank in the army, and was removed to the ROYAL NORTH BRITISH FUSILIERS in the same year. In 1772 he received the appointment of Governor of Tynemouth and Clifford's Fort; in 1777 he was promoted to the rank of lieutenant-general, and in the following year appointed Governor of Landguard Fort, from which he was afterwards removed to the government of Stirling Castle. In 1780 he was nominated Commander-in-Chief in Scotland. He died in May 1789.

THE HONOURABLE JAMES MURRAY.

Appointed 5th June 1789.

The Honourable James Murray served in the Fifteenth regiment, in which corps he attained the rank of major, and was promoted to the lieutenant-colonelcy on the 5th of January 1751. He commanded the Fifteenth in the expedition against Rochefort, under Lieutenant-General Sir John Mordaunt, in 1757, and at the capture of Louisbourg in 1758; in 1759 he commanded a brigade at the battle and capture of Quebec, under the renowned Major-General James Wolfe; in 1760 he led a division up the river St Lawrence, and contributed to the reduction of Montreal, which completed the conquest of Canada from the French. He was promoted to colonel-commandant of a battalion of the Sixtieth regiment in 1759, and to the local rank of major-general in America in 1760. In 1762 he was advanced to the rank of major-general; and in 1767 he was removed to the colonelcy of the Thirteenth regiment. He was promoted to the rank of lieutenant-general in 1772, and to that of general in 1783; in 1789 he was removed to the ROYAL NORTH BRITISH FUSILIERS. He died in 1794.

JAMES HAMILTON.

Appointed 20th June 1794.

After a progressive service in the subordinate commissions, this officer was promoted to the lieutenant-colonelcy of the TWENTY-FIRST FUSILIERS on the 11th of March 1774. He served in North America during two campaigns of the War of Independence; was promoted to the rank of major-general in 1787; and was appointed colonel of the Fifteenth foot in 1792, from which he was removed to the TWENTY-FIRST FUSILIERS in 1794. He obtained the rank of lieutenant-general in 1797, and that of general in 1802. His decease occurred in 1803.

THE HONOURABLE WILLIAM GORDON.

Appointed 6th August 1803.

The Honourable William Gordon was appointed captain in the Sixteenth Light Dragoons, when that corps was raised in the year 1759; in October 1762 he was appointed lieutenant-colonel of the One Hundred and Fifth regiment, and in 1777 he was promoted to the colonelcy of the Eighty-first regiment, which was afterwards disbanded. In 1781 he was promoted to the rank of major-general, and in 1789 was nominated colonel of the Seventy-first Highlanders. He was advanced to the rank of lieutenant-general in 1793, to that of general in 1798, and was removed to the ROYAL NORTH BRITISH FUSILIERS in 1803. He died in 1816.

JAMES, LORD FORBES.

Appointed 1st June 1816.

James, Lord Forbes, was appointed ensign in the Second Foot Guards in 1781. In 1793 he served in Flanders, under His Royal Highness the Duke of York, and commanded a company at the battle of Famars. He served at the siege of Valenciennes, and led a portion of his regiment at the storm of the outworks. He was engaged at the recapture of the post of Lincelles, where the Foot Guards distinguished themselves; also served at the siege of Dunkirk. In 1794 he served at the actions of Vaux, Cateau, Tournay, and Mouvaux; at the defence of Nimeguen and Fort St André, and in the retreat through Holland to Germany. After the action of Lincelles, in 1793, he was promoted to the rank of captain and lieutenant-colonel, in succession to Lieutenant-Colonel Bosville, who was killed on that occasion. In 1796 he obtained the rank of colonel; and in 1799 he served in the expedition to the Helder, and was present at every action of that short campaign in Holland, excepting one. In 1802 Lord Forbes was promoted to the rank of major-general, and nominated to the command of the troops stationed at Ashford, in Kent, and subsequently of the garrison at Dover, and he occasionally commanded the Kent District in the absence of Lieutenant-General Sir David Dundas and of Lord Ludlow. He was appointed second in command of the troops stationed on the island of Sicily in 1808, and promoted to the rank of lieutenant-general. On his return to England, in 1811, he was placed on the staff of Ireland.

Lord Forbes was elected one of the representative peers of Scotland in 1806, and held that distinguished situation many years. The colonelcy of the Third Garrison Battalion was conferred upon his lordship in 1806; he was removed to the Ninety-fourth regiment in 1808, to the Fifty-fourth in 1809, and to the ROYAL NORTH BRITISH FUSILIERS in 1816; in 1819 he was promoted to the rank of general. He died in 1843.

THE RIGHT HONOURABLE SIR FREDERICK ADAM, G.C.B., G.C.M.G.

Appointed 31st May 1843.

The Right Honourable Sir Frederick Adam served in the Twenty-first regiment, and, as lieutenant-colonel, commanded and led it against the French on the 17th and 18th of September 1810, on the coast of Calabria, seven miles south of Messina. He was afterwards promoted to the rank of general; and on 31st May 1843 was appointed to the colonelcy of the ROYAL NORTH BRITISH FUSILIERS. He died on the 24th of August 1853.

Sir De Lacey Evans, K.C.B.

Appointed 29th August 1853.

Sir De Lacey Evans served in India from 1807 till 1810; Portugal, Spain, and France, from 1812 till 1814; America, in 1814-1815; Belgium and France, from 1815 till 1818; and in Spain, from 1835 till 1837, as commander of the Anglo-Spanish Legion.

Present during the operations against Ameer Khan and the Pindaries, capture of the Mauritius, part of the retreat from Burgos, action on the Hormaza (wounded), battle of Vittoria, investment of Pampeluna, battle of the Pyrenees, investment of Bayonne (horse shot), actions of Vic Bijorre and Forbes, battle of Toulouse (horse shot), battle of Bladensburg (two horses shot), capture of Washington, attack on Baltimore, operations before New Orleans (boarding and capture of American flotilla), action of 25th December (wounded severely), unsuccessful assault, January (wounded severely), battle of Quatre Bras; retreat of 17th June, Waterloo (horse shot, and one sabred), investment and capitulation of Paris.

Sir De Lacey Evans received the war medal, with three clasps, for Vittoria, Pyrenees, and Toulouse; promoted to major-general 9th of November 1846, to the colonelcy of the Royal North British Fusiliers, 29th of August 1853; promoted to lieutenant-general 20th of June 1854.

He commanded the second division in the Eastern campaign of 1854, including the battles of Alma (wounded), Balaklava, and Inkerman, and siege of Sebastopol, comprising the repulse of the powerful sortie on 27th of October 1854. He continued to hold the colonelcy of the Fusiliers till his death in 1869.

Sir Frederick William Hamilton.

Appointed 11th January 1870.

Sir Frederick W. Hamilton served with the Grenadier Guards throughout the Eastern campaign, 1854-1855, including the battles of Alma, Balaklava, and Inkerman (wounded, and horse shot); sortie on 26th October, and siege of Sebastopol. During the latter part he commanded divisions of the army in the trenches, and was in temporary command of the Grenadier Guards after Inkerman. C.B. medal, with four clasps, Officer of the Legion of Honour, third class of the Medjidie, and Turkish medal. He was appointed to the colonelcy of the ROYAL NORTH BRITISH FUSILIERS, 11th January 1870, which appointment he still holds at this date—June 1885.

Appendix No. 2.

SUCCESSION OF LIEUTENANT-COLONELS

WHO HAVE COMMANDED

The Royal Scots Fusiliers,

FROM 1838 UNTIL JUNE 1885.

FIRST BATTALION.

I.

LIEUTENANT-COLONEL GEORGE DEARE.

Appointed 28th December 1838.

SUCCEEDED to the command of the Royal North British Fusiliers on the promotion of Colonel Walker, in the East Indies, in 1838; retired from the service in 1850.

II.

LIEUTENANT-COLONEL JOHN CROFTON PEDDIE.

Appointed in 1850,

In succession to Colonel Deare; exchanged into the Forty-first regiment, with Lieutenant-Colonel Thomas Gore Browne, 1850.

III.

Lieutenant-Colonel Thomas Gore Browne.

Appointed in 1850,

In exchange with Colonel Peddie; retired in 1851.

IV.

Lieutenant-Colonel Edward Thorpe.

Appointed in 1851,

From Eighty-ninth regiment; retired in 1852.

V.

Lieutenant-Colonel Frederick George Ainslie.

Appointed in 1852,

From senior regimental major; died at Scutari, 14th of November 1854, of wounds received at the battle of Inkerman.

VI.

Lieutenant-Colonel John Ramsay Stuart.

Appointed 15th November 1854.

Senior regimental major, in succession to Colonel Ainslie. Served in the Crimean War of 1854, and the latter part of 1855, including the battles of Alma, Balaklava, Inkerman, siege and fall of Sebastopol; twice mentioned in despatches. Medal with four clasps, fifth class of the Medjidie, Turkish medal, and Companion of the Bath. Retired from the command on the 17th of April 1867; promoted major-general 6th March 1868, and held the command of the troops in the North British district until his promotion to lieutenant-general. Is colonel of the Dorset regiment. Promoted to general 3d December 1880, and is now on the retired list.

VII.

Lieutenant-Colonel John Thomas Dalyell.

Appointed 17th April 1867,

From senior regimental major; served in the Eastern campaign of 1854, and part of 1855, including battles of Alma, Inkerman, and siege of Sebastopol; brevet-major, medal with three clasps, fifth class of the Medjidie, and Turkish medal; retired from the command in January 1878, on obtaining the command of the First Regimental District, at Glencorse; promoted to major-general, on the active list, on the 27th June 1883.

VIII.

Lieutenant-Colonel Alfred Templeman.

Appointed 23d January 1878,

From senior regimental major; served throughout the Eastern campaign, 1854-55, including the battles of Alma, Balaklava, and Inkerman (wounded), siege and fall of Sebastopol, and attack on the Great Redan, on the 18th June 1855, also the expedition to Kinburn. Medal with four clasps, Knight of the French Legion of Honour, and Turkish medal. Retired from the command on the completion of the regulated period of five years, in 1882. Subsequently appointed to the command of the Ninety-first Regimental District, at Stirling Castle, where he still remains (June 1885).

IX.

Lieutenant-Colonel George Frederick Gildea.

Appointed 1879,

To second battalion, in succession to Colonel W. P. Collingwood; subsequently exchanged to the first battalion with Lieutenant-Colonel Winsloe.

Served as adjutant of the Eleventh Turkish contingent from September 1855, to its disbandment in May 1856; Turkish medal.

Commanded the second battalion, Royal Scots Fusiliers, and the garrison of Pretoria, during the Transvaal War of 1881; commanded a force in several engagements with the Boers and sorties made from Pretoria; severely wounded on the 12th February; mentioned in despatches. Medal and clasp for South Africa, and appointed A.D.C. to the Queen.

Retired from the command on the 24th of August 1884, on the completion of the regulated period of five years; subsequently appointed to the staff of the army as assistant-adjutant and quartermaster-general. Served on the staff of Major-General Graham at Suakim; but, from over-exertion in the performance of his duties, his health failed, and he was invalided to England.

X.

Lieutenant-Colonel Edward Thomas Bainbridge.

Appointed 24th August 1884,

From senior regimental major. Served with the second battalion, Royal Scots Fusiliers, in the "Zulu campaign" of 1879; was engaged in the operations against Sekukuni, including the storming of his stronghold, where he commanded the storming party; mentioned in despatches. Medal with clasp.

Lieutenant-Colonel Bainbridge still retains command of the first battalion, Royal Scots Fusiliers, at Portland (June 1885).

SECOND BATTALION.

I.

LIEUTENANT-COLONEL EDWARD LAST.

Promoted from the Ninety-ninth regiment in 1858, on the formation of the second battalion, Royal Scots Fusiliers; retired from the service in 1859.

II.

LIEUTENANT-COLONEL E. LOWE.

Appointed September 1859.

Promoted from the Thirty-second Light Infantry; exchanged in June 1862 with Colonel J. G. Robertson, from the Sixth Foot.

III.

LIEUTENANT-COLONEL J. G. ROBERTSON.

Appointed June 1862,

On transfer from Sixth regiment; transferred to the staff of the Indian army, and appointed adjutant-general, Madras Presidency.

IV.

LIEUTENANT-COLONEL E. A. T. STEWARD.

Appointed October 1868,

From senior major of the regiment; served in the Crimean campaign of 1854-55 with the first battalion, Royal Scots Fusiliers, including the battles of Alma, Balaklava, Inkerman, and the siege of Sebastopol. Medal with four clasps, and Turkish medal. Retired from the service on account of ill-health in 1870.

V.

Lieutenant-Colonel Frederick Lyster.

Appointed in 1870,

From senior major of the regiment; served with the Fiftieth regiment in the battle of "Punniar;" bronze star. Previous to entering the British army, Colonel Lyster had seen much service under Don Pedro, in Portugal, from 1833 to 1835. Retired from the command of the battalion in 1874.

VI.

Lieutenant-Colonel William Pole Collingwood.

Appointed 23d May 1874,

From senior major of the regiment; served with the Thirty-seventh regiment in the Ceylon Rebellion of 1848; he was present with the first battalion, Royal Scots Fusiliers, in the Crimea, from 11th July 1855 to the close of the campaign, including latter part of siege, fall of Sebastopol, and expedition to Kinburn; from November 1855 to November 1856, he commanded a battalion of the land transport; Crimean medal with one clasp, and Turkish medal. Proceeded to South Africa in command of battalion; commanded second brigade of second division of the army during the latter part of the Zulu War; medal with clasp, and C.M.G. Retired from the command after the completion of the regulated period; subsequently appointed to the command of the Fifth Regimental District, Newcastle-upon-Tyne, where he is at present (June 1885).

VII.

Lieutenant-Colonel George Frederick Gildea.

Appointed 1879,

From senior major of the regiment; exchanged to first battalion command with Colonel Winsloe. (For services, see Commanding Officers, First Battalion, No. IX.)

VIII.

Lieutenant-Colonel Richard William Charles Winsloe.

Appointed February 1882,

From senior major of the regiment, and exchanged with Lieutenant-Colonel Gildea to first battalion. Served with the first battalion, Royal Scots Fusiliers, in the Crimea, from 6th August 1855, including latter part of siege, fall of Sebastopol, and expedition to Kinburn. Medal and clasp, and Turkish medal.

He also served with the second battalion, Royal Scots Fusiliers, in the Zulu campaign, including the battle of Ulundi (wounded severely); also in the Transvaal campaign. Commanded the troops at Potchefstroom during the gallant and determined defence of the fort at that place. Medal and clasps for South Africa, and appointed A.D.C. to the Queen. Colonel Winsloe commands the battalion at Burmah at this date (June 1885).

Appendix No. 3.

LIST of BATTLES, SIEGES, &c., in the NETHERLANDS during the Reign of KING WILLIAM III., from 1689 to the Peace of Ryswick in 1697.

Battle of Walcourt	25	August	1689
Battle of Fleurus	1	July	1690
Mons surrendered to the French	10	April	1691
Namur surrendered to the French . . .	20	June	1692
Battle of Steenkirk	3	August	1692
Furnes and Dixmude captured	—	Sept.	1692
The French lines at D'Otignies forced . .	10	July	1693
Battle of Landen	29	July	1693
Surrender of Huy	17	Sept.	1694
Attack on Fort Kenoque	9	June	1695
Dixmude surrendered to the French . . .	16	July	1695
Deinse surrendered to the French . . .	21	July	1695
Namur retaken by King William III. . .	25	July	1695
Citadel of Namur surrendered	5	Sept.	1695
Treaty of Ryswick signed	11	Sept.	1697

LIST of SIEGES, BATTLES, &c., in the NETHERLANDS and GERMANY during the Campaigns under the DUKE OF MARLBOROUGH, from 1702 to 1711.

	Invested.	Surrendered.	
Siege of Kayserswerth	16 April	17 June	1702
Skirmish near Nimeguen		11 June	1702
Siege of Venloo	29 Aug.	25 Sept.	1702
Capture of Fort St Michael		18 Sept.	1702
Siege of Stevenswaert		3 Oct.	1702
Siege of Ruremonde		6 Oct.	1702
Capture of Liege Citadel		23 Oct.	1702
Siege of Bonn	24 April	15 May	1703
Siege of Huy	16 Aug.	25 Aug.	1703
Siege of Limburg	10 Sept.	28 Sept.	1703
Battle of Schellenberg		2 July	1704
Battle of Blenheim		13 Aug.	1704
Siege of Landau	12 Sept.	24 Nov.	1704
Huy captured by the French		May	1705
Recapture of Huy		11 July	1705
Forcing the French Lines at Helixem, near Tirlemont		18 July	1705
Skirmish near the Dyle		21 July	1705
Siege of Sandvliet	26 Oct.	29 Oct.	1705
Battle of Ramillies		23 May	1706
Siege of Ostend	28 June	8 July	1706
Siege of Menin	25 July	25 Aug.	1706
Siege of Dendermond	29 Aug.	5 Sept.	1706
Siege of Aeth	16 Sept.	3 Oct.	1706
Battle of Oudenarde		11 July	1708
Siege of Lisle	13 Aug.	23 Oct.	1708
Capture of the Citadel		9 Dec.	1708
Battle of Wynendale		28 Sept.	1708
Passage of the Scheldt		27 Nov.	1708
Siege of Ghent	18 Dec.	30 Dec.	1708
Siege of Tournay	27 June	29 July	1709
Capture of the Citadel		3 Sept.	1709
Battle of Malplaquet		11 Sept.	1709
Siege of Mons	21 Sept.	20 Oct.	1709
Passage of the French Lines at Pont à Vendin		21 April	1710
Siege of Douay	25 April	27 June	1710
Siege of Bethune	15 July	29 Aug.	1710
Siege of Aire	6 Sept.	9 Nov.	1710
Siege of St Venant	6 Sept.	30 Sept.	1710
Passage of the French Lines at Arleux		5 Aug.	1711
Siege of Bouchain	10 Aug.	13 Sept.	1711
Treaty of Utrecht signed		30 March	1713

BATTLES, SIEGES, &c., which occurred in GERMANY and in the NETHERLANDS from 1743 to 1748, during the "War of the Austrian Succession."

Battle of Dettingen (Germany)	27 June	1743
Menin invested by the French 18 May, and captured	5 June	1744
Ypres invested by the French 7 June, and captured	14 June	1744
Fort Kenoque surrendered to the French . .	June	1744
Furnes surrendered to the French	5 July	1744
Friburg (Germany) invested by the French . .	21 Sept.	1744
Citadel of Friburg captured by the French . .	28 Nov.	1744
Tournay invested by the French	23 April	1745
Battle of Fontenoy	11 May	1745
Citadel of Tournay surrendered to the French .	21 June	1745
Skirmish near Ghent	9 July	1745
Ghent captured by the French	30 June	1745
Bruges captured by the French	July	1745
Oudenarde captured by the French	21 July	1745
Dendermond captured by the French . . .	Aug.	1745
Ostend invested by the French on 14 July, and captured	23 Aug.	1745
Nieuport captured by the French	26 Aug.	1745
Aeth captured by the French	28 Sept.	1745
Brussels invested by the French 24 Jan., and captured	20 Feb.	1746
Mechlin captured by the French	May	1746
Antwerp captured by the French	20 May	1746
Citadel of Antwerp captured by the French . .	31 May	1746
Mons invested by the French on 7 June, and captured	11 July	1746
Fort St Ghislain captured by the French . .	21 July	1746
Charleroi invested by the French on 25 July, and captured	2 Aug.	1746
Huy captured by the French	Aug.	1746
Namur invested by the French 26 Aug., and citadel captured	19 Sept.	1746
Battle of Roucoux, near Liege	11 Oct.	1746
Sluys surrendered to the French	11 April	1747
Fort Sandberg, in Hulst, and Axel surrendered to the French	28 April	1747
Sandvliet captured by the French	June	1747
Battle of Val, or Laffeld, near Maestricht . .	2 July	1747
Bergen-op-Zoom invested by the French 13 July, and captured	16 Sept.	1747
Forts Lillo, Frederick Henry, and Croix . .	2 Oct.	1747
Limburg captured by the French	March	1748
Maestricht invested by the French	3 April	1748
Maestricht surrendered to the French	3 May	1748
Treaty of Aix-la-Chapelle	7 Oct.	1748

LIST of the BRITISH REGIMENTS which served in Flanders and Germany between 1742 and 1748, during the "War of the Austrian Succession."

REGIMENTS.		Year in which embarked for Flanders.	Returned to Great Britain in consequence of the Rebellion in favour of the Pretender.	Rejoined the Army in Flanders after the suppression of the Rebellion.
CAVALRY.	COLONELS.			
3d Troop Horse Guards	Earl of Albemarle ...	1742	1746	...
4th ditto ditto	Earl of Effingham ...	1742	1746	...
2d ditto Horse Grenadier Guards ...	Earl of Craufurd......	1742	1746	...
Royal Regiment Horse Guards ...	Earl of Hertford ...	1742	1746	...
1st Horse (1st Dragoon Guards)......	Earl of Pembroke ...	1742	1746	...
4th Irish Horse (7th Dragoon Guards)	Sir John Ligonier ...	1742	1746	...
1st Dragoons	Hawley	1742	1746	...
2d ditto	Campbell	1742	Rem'd in Flanders	...
3d ditto	Honeywood	1742	1746	...
4th ditto	Rich	1742	1746	1747
6th ditto (Inniskilling)	Lord Cadogan	1742	Remained in Flandars	...
7th ditto	Cope	1742		
FOOT GUARDS.				
1st Foot Guards, 1st Batt.	Duke of Cumberland	1742	1746	1747
2d ditto ditto.	Duke of Marlborough	1742	1746	...
3d ditto ditto.	Earl of Dunmore	1742	1746	1747
INFANTRY.				
1st Foot, 1st Battalion	St Clair	1744	1746	...
3d ditto (Buffs)	Howard	1742	1746	1747
4th ditto	Barrel	1744	1746	...
8th ditto	Onslow	1742	1746	1747
11th ditto	Cornwallis	1742	Rem'd in Flanders	...
12th ditto	Duroure	1742	1746	...
13th ditto	Pulteney...............	1742	1746	1747
18th ditto	Mordaunt	1743	1746	...
19th ditto (Green)......	Howard	1744	Rem'd in Flanders	...
20th ditto	Bligh	1742	1746	1747
21st ditto, Royal Nth. British Fusiliers...	Campbell	1742	1746	1747
23d ditto, Royal Welsh Fusiliers ...	Peers	1742	1746	1747
25th ditto	Earl of Rothes	1744	1746	1747
28th ditto	Bragg	1744	1746	...
31st ditto	Handasyd	1742	1746	...
32d ditto	Skelton	1742	Remained in Flanders	...
33d ditto	Johnson	1742		
34th ditto	Cholmondeley	1744	1746	...
36th ditto	Fleming	1744	1746	1747
37th ditto	Ponsonby	1742	1746	1747
42d ditto	Lord Semphill	1744	1746	...
48th ditto	Lord Harry Beauclerk	1744	1746	1747

Appendix No. 4.

Additional Distinctions or Decorations specially conferred upon Officers, Non-Commissioned Officers, and Men of the First Battalion Royal Scots Fusiliers, for services during the Crimean Campaign of 1854-1855, other than the Crimean Medal, with clasps, and Turkish War Medal, issued to all present with the Army.

Knights of the French Legion of Honour.

To Captain Alfred Templeman, for uninterrupted service throughout the campaign, and steady performance of the arduous trench work during the siege; served with the Twenty-first Fusiliers throughout the Eastern campaign of 1854-1855, including the battles of Alma, Balaklava, Inkerman (wounded), siege and fall of Sebastopol, and expedition to Kinburn. Crimean medal, with four clasps, and the Turkish medal.

To Captain J. G. Image, for uninterrupted service throughout the campaign, including the battles of Alma, Balaklava, Inkerman, on ladder party at the attack on the Redan on 18th June, siege and fall of Sebastopol, and capture of Kinburn, Medal, with four clasps, and Turkish medal. This officer died in 1870 in Bangalore, East Indies.

To Captain Roger Killeen, for serving throughout the campaign of 1854-1855, including battles of Alma, siege and fall of Sebastopol, Balaklava, Inkerman, attack on the Redan on 18th June, and capture of Kinburn. At Inkerman he commanded a company, and, under a heavy fire, rescued the regimental colour; he was severely wounded on the occasion, but did not quit the field, although directed to do so. Medal, with four clasps, and Turkish medal. This officer is now dead.

The French War Medal.

To Colour-Sergeant John Higdon, for distinguished conduct at the battle of Inkerman. This non-commissioned officer obtained his discharge, and is since dead.

To Private Timothy Driscoll, for uninterrupted service in the trenches throughout the siege.

To Sergeant Patrick Kelly, for uninterrupted service in the trenches throughout the siege. This non-commissioned officer subsequently obtained a commission in the First West India regiment as an ensign, and was killed in storming a stockade on the West Coast of Africa.

To Private Michael M'Phely, for conspicuous bravery in the sortie of 23d March 1855, when the Russians were beaten back with great loss, and for general gallant conduct throughout the siege.

To Sergeant John Russell, for gallant conduct at the battle of Inkerman, and uninterrupted performance of duties throughout the siege.

To Private Peter Crowley, for distinguished conduct in rescuing an officer of another regiment from a Russian soldier, who had taken him prisoner, at the battle of Inkerman.

To Colour-Sergeant Richard Ellis; present at Alma, wounded at Inkerman.

To Sergeant Edward Marshman, for continuous service throughout the campaign, as an active zealous non-commissioned officer. He volunteered for a sortie, and led a party against the Russians, for which service he was promoted to corporal, and subsequently to sergeant.

To Sergeant James Sim, for gallant conduct, on the 18th of June 1855, when on the ladder party in the attack on the Redan. Medal and four clasps, and Turkish medal. Severely wounded. This non-commissioned officer afterwards obtained a commission, and is now camp quartermaster at Aldershot.

The Sardinian War Medal.

To Colonel C. R. S. West, Viscount Canteloupe, and afterwards Earl De la Warre, for service throughout the whole campaign of 1854-1855 ; present at Alma and Inkerman ; commanded the left wing of the regiment at the latter, and the whole regiment in the attacks on the Redan on the 18th of June and 8th September 1855 ; also at the capture of Kinburn. This officer is since dead.

To Brevet-Major George N. Boldero, for service in campaign of 1854, including battles of Alma and Inkerman, at the latter of which he was severely wounded ; subsequent siege of Sebastopol. Now major-general.

To Captain H. King, for service with regiment from landing on 14th of September 1854 ; present at battles of Alma and Inkerman. Dangerously wounded at the latter, being shot through the neck. Drowned whilst bathing at Malta.

To Captain Richard Stephens, for service with the regiment from date of landing in the Crimea, including the battles of Alma and Inkerman. At the latter he lost an arm while carrying the Queen's colour. He performed trench duty up to the 5th of November 1854, until invalided to England from wounds.

To Captain S. H. Clerke, for service from 14th September 1854 to 26th October 1854, and from 27th December 1854 to end of war ; present at the battle of Alma and expedition to Kinburn. At present lieutenant-colonel on the retired list, and in Her Majesty's Bodyguard.

To Sergeant-Major Fowler, for service throughout the campaign ; present at the battles of Alma, Balaklava, and Inkerman. Subsequently he obtained a commission as ensign in the 4th West India regiment ; took part in the Ashantee campaign ; and died, on half-pay as lieutenant-colonel, from the effects of illness contracted on African service.

To Colour-Sergeant Richard Ellis ; served with his regiment until ordered to join reserve battalion at Malta on

11th May 1855; transferred to second battalion Third regiment as sergeant-major.

The following non-commissioned officer and men received the silver Medal for Distinguished Conduct in the Field:—*Colour-Sergeant William Rogers, Private Patrick Murray, Private Patrick M'Guire, Private Murtiff Maddigan.*

Specially mentioned in General Lord Wolseley's despatch of the Soudan Campaign.

Royal Scots Fusiliers, Second Battalion—*Sergeant Duggan*, employed with the Mounted Infantry.

Appendix No. 5.

LONG, LONG AGO!

In concluding the historical record of the Royal Scots Fusiliers, it may not be out of place to add a few notes as to the position, dress, equipment, and life of a soldier fifty years ago.

The recruits who joined the headquarters of their regiments, or Four Company Depots, direct, were more fortunate as compared with those who had enlisted for corps serving in India, and whose depots were stationed with the provisional battalion at Chatham, vulgarly known as the "Pongoes."

In this establishment, every advantage was taken of the ignorance and inexperience of the recruit; he was starved and robbed with impunity, as he either did not know that he had any means of redress, or was afraid to seek it. However it happened, the pay-sergeants in the Chatham garrison were adepts at bamboozling recruits, and perfect masters of chicanery.

A most nefarious practice existed of issuing to each man old and unserviceable "firelocks;" the recruit, in his ignorance, was unable to detect defects or deficiencies, but in a week or so afterwards, when returning them to the armourer sergeant, previous to embarkation for India, these defects and deficiencies were painfully brought under the unfortunate's notice, in the form of heavy charges to make them good. The armourer had a list of the arms, and by referring to the number on the paper, and comparing it with that on the firelock, *he knew where to look for the defect.* How many times over these reported damages had been paid for, *but never repaired*, is a mystery only known to the armourer and pay-sergeants.

On Thursdays and Sundays the men were compelled to have what was called "baked dinners;" but there being no bake-houses in the barracks, the dinners had to be sent to Brompton to be cooked. The original quantity was very small for the number who had to partake of it, but it was very much smaller when returned. On these days hunger was rather increased than appeased; men have risen from their dinner with the cravings of hunger stronger than when they sat down, with no prospect of food for another nineteen hours.

Men wore padding in the breasts of their clothing in those days, and there was a sarcastic proverb in vogue to the effect that "*Soldiers had full breasts and empty stomachs.*"

In the orderly room of this garrison the letter-box was kept; all letters had to be sent in open, that is unsealed or unwafered, envelopes being unknown. They were "franked" by the commanding-officer, which exempted them from postage; but it did not exempt the recruit from being fleeced the sum of one penny as the drum-major's fee for posting it. A similar payment was demanded for all letters received, so that in some instances it happened that a letter would be brought to a man who had not the penny to pay for it; in this case it would be retained by the drum-major until his fee was paid. A trying ordeal for the young lad eager and anxious to hear from home!

The soldier's pay was at that time one shilling per diem, and one penny, called "beer money," in lieu of a "spirit ration," when not issued in kind. The period of service was for twenty-one years, and latterly for unlimited service, commencing at the age of eighteen; any service under that not counting for pension.

Additional pay for length of service, of twopence after fourteen years. No good-conduct pay was granted until December 1845.

On enlistment a bounty of from £3 to £4 was given. This was theoretically supposed to meet all the expense of kit and outfit, but practically such was not the case, as with one item and another it cost upwards of £5. 10s.; consequently the unfortunate recruit was in debt for the first six months of his service, receiving the munificent sum of one penny per diem to provide cleaning requisites and pocket money.

Pension from sixpence to tenpence per diem. Out of his daily pay and beer money of one shilling and a penny per diem, he had to defray the cost of his rations, messing, and washing, which came to eightpence per diem; and with the renewal of necessaries, haircutting, marking, sheet-washing, and barrack damages, the average daily charge amounted to about tenpence halfpenny,—rather a large slice out of one shilling and a penny.

Only two meals per diem were provided, these scanty and inferior; and unless poor Tommy Atkins laid by a portion of his dinner, he went supperless to bed.

The "brown tommy" issued for breakfast was so dense and doughy, that the regulated ration of one pound was of a very small size. It was commonly believed, that if thrown against a wall it would adhere to it; it was never known that the experiment was tried, as every man had another and more urgent use for it.

The soldier had also to pay for browning his gun barrel, alteration of clothing, pipe-clay, cleaning arms, and, in the winter months, for coal to heat his barrack-room, the Government allowance being insufficient.

No schools, except for children, volunteer pupils, and non-commissioned officers, for which the latter paid a monthly charge of tenpence.

No library or recreation rooms.

No musketry prizes.

The establishment of a regimental savings bank in those days would have been cruel irony; it can be quite understood, that with only a net daily pay of twopence halfpenny, it would have taxed "Daniel Dancer" to effect a deposit. We read that even the great "Cobbett" had, when a private soldier, to practise the severest economy to provide a dip candle, to carry out his self-imposed task of education, and only occasionally was able to enjoy the luxury of a "red herring" for breakfast.

Canteens were leased by civilian tenants; goods of the worst description were sold at the highest prices.

Clothing, only one suit per annum, viz.:—Coatee, one pair of cloth trousers, and one pair of boots; shell jacket, fatigue and white trousers, and second pair of boots, paid for by the soldier.

The "ammunition" boots issued were so bad that after being worn half a dozen times they often fell to pieces, and had to be "made over," *i.e.*, entirely renewed, at a cost of from six shillings and sixpence to ten shillings, according to circumstances. The clothing was very badly made; to furnish extra fringe for "wings" (not to fly with), grenades for collars, skirts, and shells, padding, lining, and extra lace, the private soldier had to pay an annual charge of about four shillings and sixpence for alteration and fitting of clothing, sergeants more in proportion.

Greatcoats were allowed to be worn only on duty, and kept in repair at the expense of the soldier.

Parades three times a day, one at least in "heavy marching order," except in winter, when only two per diem took place.

A soldier's time was almost wholly taken up in cleaning arms, accoutrements, clothing, and barrack-room "shining," or on fatigue; no leisure time; to be on guard was a rest. A modern soldier has no conception what was meant by a clean barrack-room in former times; cots correctly dressed at all hours, bedding folded with a neatness and precision now totally unknown; everything on the shelves folded with the same neatness, and nothing allowed to appear but what was strictly regimental; not even a looking-glass was permitted. The floors were scrubbed cleaner than modern tables; a drop of pipe-clay on the window-sill represented three days confined to barracks for the "orderly man;" and woe betide the thoughtless individual who stood upon the well-whitened hearthstone! Lying down or lounging in a barrack-room was entirely unknown; the bedding was brought forward to the edge of the cot, where it remained until retreat beating, before which the beds were not permitted to be made down; the only mode of relieving the upright position was to sit on a form and lean back against the edge of a table.

A Fusilier's "coatee," with "wings," required three or four hours for thoroughly cleaning and preparing it for guard mounting.

The preparation of the pouch was an art in itself, which very few properly understood. What with scraping, heel-balling, and bees-wax polishing, it was the greatest care poor Tommy had upon his mind; and when properly polished, and ready for parade, was an excellent substitute for a looking-glass. To preserve it in this precious state of polished perfection, a waterproof cover, carefully lined with the finest linen, had to be used.

The locks of the old "Brown Bess" were bright in those good old days; and on Saturday, at "kit inspection," each man paraded with his burnished lock in one hand, and "his pouch" in the other. For the purpose of drill a bone "snapper" was used instead of a flint; the snapper, or flint, was placed between two flat pieces of lead, with milled edges, called "leads,"—those also had to be always bright and shining.

The rolling of the greatcoat for heavy marching order parades, or folding it for guard mounting, were operations of the greatest difficulty, and required the aid of comrades to perform. The ends of the rolled coat had to be made perfectly even, by the aid of a fork and a spoon; bulging parts were pushed into their places by the handle of the spoon, and the depressed parts brought out to a level by aid of the fork; the operation was completed by sewing the skirt end firmly to the roll. The folded coat was not considered perfect until it had been well damped, and pressed under a heavy iron coal box all night, to give it a thin compact appearance. All this preparation for guard mounting was to secure the proud position of being selected as commanding-officer's orderly for a day; the cleanest and smartest man being always given this post, which secured the benefit of exemption from sentry go, and the night in bed. To such an extent was emulation for this distinction carried out, it is known as a fact that well rolled or folded great-coats were held at a premium, and passed from hand to hand. A wet day was a great calamity, and the order to parade in loose coats created dismay amongst the owners of these valued productions of the folding or rolling art.

To save the full-dress clothing, provided by the head colonel, shell jackets were worn on every possible occasion; for these the men paid from ten to thirteen shillings each. Coatees were held sacred for state occasions, church parades, guard mounting, and commanding-officer's parade. White trousers were worn in summer, each man had two pairs; these were washed and scrubbed by the soldier, and it was not unusual for them to be worn in a half-dried state.

Latterly these were replaced by a pair of lavender-coloured woollen ones; and as each captain had his own opinion about the particular shade of colour they should be,—sky-blue, sea-blue, or slate-blue,—various mixtures had to be prepared by Tommy to meet this new and varied artistic taste.

One, and not the least, of "Tommy's" miseries, was the frequent and long marches he had to undertake, when moving from one station to another, whether in summer's heat or winter's storm.

Railroads did not exist in those days, so the soldier had to trudge along his weary way, over twenty miles of bad road each day, often with blistered feet. Neither "kit bags" nor "squad bags" were thought of. Poor Tommy had to carry all his worldly possessions on his back, like the snail; loaded like a donkey; scarcely able to breathe, from the pressure of the *shoulder and breast straps* of that now obsolete and murderous article of military equipment, "*the knapsack with straps complete.*"

The sergeants' pay was one shilling and tenpence; after paying from tenpence to one shilling for messing, five shillings a month to a "batman," two shillings and sixpence for washing, and other incidental charges, he was worse off than the private. He was compelled to be always well dressed and smart in appearance, which entailed expensive shell jackets, trousers, boots, cap, and gloves. When it is stated that a sergeant's jacket cost about £2, and trousers £1, and that he had to provide one or more of each every year, it can be easily understood that his financial position was not a flourishing one.

Married soldiers were merely tolerated; no provision made for their accommodation beyond the corner of a barrack-room with the single men,—living quite in public amongst them, the space being only allowed to be screened off during the night; latterly this was changed, *and, as a great improvement, five or six* married families were allowed to occupy *one room*, originally intended for the accommodation of twelve men. There was no separation allowance, fuel or light, free half-rations, or money payment of any kind, nothing to eke out an existence but what could be earned at the wash-tub.

Tattoo was at 8.30 P.M. in winter, and at 9.30 P.M. in summer; passes were withheld for the slightest offence, furloughs granted sparingly, and men placed in the "guard-room" for the most trivial crime.

Punishments were long, heavy, and severe, ranging from three days' drill by the captains of companies to fourteen by the commanding-officer. A peculiar system of punishment existed called the "dry room," which meant separation from comrades, several drills daily in heavy marching order and fatigues. Corporal punishment was common, varying from 150 to 1000 lashes, and inflicted for very trifling offences.

The Uniform and Equipment of Long, Long Ago!

The title "Fusiliers," or "Fuzileers," was given to certain regiments armed with a musket made especially short, to be carried conveniently slung over the shoulder, so as to leave the hands free for other purposes.

The officers carried "spontoons," and sergeants "pikes"

and "halberts." Subsequently the spontoon of the officer was changed for the sword, and the pike or halbert of the sergeant for the "fusil," or more commonly known "fusee." This again has been changed, as sergeants now use the same description of rifles as the men, the only distinction being that they carry "sword bayonets," while the men still retain the well-known "fork with one prong," *i.e.*, the ordinary bayonet.

In 1832—previous to the breaking up of the regiment into small detachments for duty on board convict ships proceeding to Van Dieman's Land, now known as the island of Tasmania—it had reached, for those days, the nearest possible approach to perfection as a military organisation. Stalwart men, from five feet seven inches to six feet four inches in height; perfectly drilled in the rigid manœuvres of the period; and so steady, that such a thing as a wink of the eyelid, or a sneeze, while in the ranks, would be ruinous to the offender.

Skin-tight swallow-tailed coatees were worn, trimmed with white lace; white metal (leaden) buttons; semi-circular erections round the top of each shoulder called "shells," covered with cloth and lace, and finished off with several rows of white lambs-wool fringe called "wings;" long skirts, ornamented with brass grenades, with another on each side of the collar, completed this garment. Then there was the never-to-be-forgotten military article of torture, "*the black leather stock with clasps complete*," which had to be worn continually. To become familiarised with it men have been known to sleep with it at night; it was at meals that its inconvenience was most felt, more especially by the young recruit, who, to be able to see his dinner, had to place his plate at arm's length from him on the table. The stock gradually decreased in depth and thickness, and at last, to the great joy of the soldier, totally disappeared.

Blue trousers, with a broad scarlet stripe, and wide mouth bottoms *à la* "Jack Tar."

Lace-up boots, with narrow toes, high heels, and very much polished.

Broad buff belts worn over each shoulder, and crossing the breast; at the intersection of these was affixed the brass "breast-plate," bearing upon it the regimental badge, a raised and movable grenade with the motto of the House of Hanover round its ball, "*Nec aspera terrent*," and the number XXI. in the centre. To bind the cross-belts and keep the accoutrements steady a narrow waist-belt was worn, which was fastened by a small plate bearing as its device the "*Thistle*," with its motto "*Nemo me impune lacessit*." From one of the lower corners of the breast-plate hung a narrow buff strap, divided at the lower end into two parts, to which were attached the "*picker and brush*," so necessary to keep old "Brown Bess" in order,—the picker to keep the touch-hole clear, and the brush to clean the priming-pan when fouled by frequent firing.

Attached behind to one of the cross belts was "the pouch," a thing to wonder at, nearly as large as a "Gladstone bag," and capable of holding sixty rounds of "ball ammunition," with a supply of "flints," "turn-screw and worm." But, alas, in those days of degenerate "peace with honour," ammunition was only carried to the extent of ten rounds per man; the remaining spaces of the pouch being filled with "blanks," or "dummies," made to represent packets of ammunition, and neatly marked on the outside with the owner's name, company, and regimental number.

The polish and appearance of this pouch, and the manipulation required to bring it to a perfect state, have already been noticed; on its flap was placed a large brass grenade, mounted on scarlet cloth, and polished to a painful degree of brassy brightness.

This military uniform of the period was completed by the head-dress, a large black bearskin cap, similar to those formerly worn by the Foot Guards, a white upright feather "hackle" in a grenade socket at the right side, tapering brass "scales" as a chin-strap, and several thick braids of white cord passing around it and fastened to the left shoulder strap button, similar to the "life lines" now worn by some of the cavalry corps.

The arms were the simple "Brown Bess" flint musket of the day. The stock was a matter of great pride and anxiety, being sometimes artificially striped like a tiger's skin, French polished, or bees-waxed; it had to be handled most carefully to avoid scratches. The bayonets were fixed on the muzzle with a slit, but without a spring or ring; and after the "manual exercise" had been completed, an amusing series of gymnastic performances took place, in recovering as quickly as possible the bayonets which had flown off or stuck in the ground when obeying the command "charge bayonets." The brass pins securing the barrel to the stock were by the soldier loosened, in order that when performing the "manual exercise" the "motion" should "tell," *i.e.*, make the loudest possible noise.

The sergeants carried short "fusils," or "fusees," and were also armed with both sword and bayonet; they wore a blue and scarlet striped sash round their waists, and from the right side of it two large tassels like bell-pulls were suspended.

The band wore "blue" coatees, with scarlet epaulettes, aguilettes, and facings, a red feather "hackle" in their bear-skin caps, light blue trousers with a double scarlet stripe, swords of a scimitar shape with brass scabbards. As usual the "*Big Drummer*" was a man of colour from the West Indies, a splendid specimen of a well-grown negro. The "*Jingling-Johnny*," a peculiar instrument on a long pole, covered with bells, crescents, and horsehair, was carried in front of the big drum by the tallest man in the regiment.

The walking drum-major was a sight worth seeing; generally one of the handsomest and best-proportioned men in the regiment, whose duties were to wield his staff in the most wonderful manner, and keep his walking-cane in order by frequent use of it on refractory drum boys.

The major's dress was gorgeous,—a magnificent and perfectly fitting coatee, bedizened with silver lace, plated buttons, silver lace wings and fringe; aguilettes; richly embroidered shoulder belt, with regimental badges, honours, and insignia in gold; Hessian boots with tassels; tight fitting knee-breeches, or pantaloons; an extra large-sized bearskin cap, with silver cord life lines attached, and at its right side a gallant scarlet upright feather plume; long white buckskin gauntlets; and in his hand a handsome massive staff of Malacca cane, surmounted by a gilt grenade, and a rich cord of gold encircling it from top to bottom.

There is a story told of one poor drum-major, who outgrowing the symmetry of his beautiful figure, his coat became too tight. Horror! what was to be done? Starvation the drum-major resented, neither would he diminish the quantity of his beer. A pair of stays were made for him, into which two drummers managed to lace him; but the major became stouter, the lacing became tighter, and in the end he went to hospital and died, a victim to "drum-major dandyism."

During the station of the regiment in Van Dieman's Land it discarded its drums and fifes, and in lieu thereof established a key-bugle band, which remained in existence for many years.

In 1857, when at Malta, the drummers were formed into a brass band, and proved of great service in supplementing the band when playing out.

It also maintained two Highland pipers up till the year 1850; these were again resumed, and the number increased to ten, in 1870.

The drummers of old were gay fellows, and "great little swells;" with their coatees bedizened with variegated lace, large shoulder wings to match, bearskin caps on their heads nearly as high as themselves, carrying drums equally out of proportion, buff knee aprons, swords and belts, drum carriages with brass plate rollers to hold the drumsticks, and large bugles in their hands, they indeed made a brave appearance.

The Fusiliers, when inspected at Windsor by the Duke of Clarence, afterwards William the Fourth, had a more splendid appearance than the Foot Guards, in consequence of which he caused an order to be issued from the Horse Guards to discontinue the use of all articles not strictly regulation.

Regulations affecting the dress and appearance of a regiment cannot be carried out very promptly, as time must be allowed to wear out articles in use before others of the substituted pattern can be issued ; it was, therefore, some years before these changes in the uniform were really carried into effect, as the regiment continued to wear the same style of dress until 1838, when the bearskin caps were replaced by chakos of the "bell-top pattern," with an upright white hair hackle.

This head-dress, and succeeding ones, except the seal-skin cap or Fusilier busby now in use, was known by the peculiar double title of "hat cap," a term used for the same article in the time of Queen Elizabeth.

New accoutrements were issued to the regiment, and the pouch ornament discontinued, in 1842. In 1846 the flint musket was replaced by the percussion one, and pickers and brushes discontinued.

The band was clothed in white instead of blue; the colour of the trousers was also changed from blue to an Oxford grey, with a narrow red cloth welt in lieu of the broad scarlet stripe.

In 1841 the "bell-topped chako" was replaced by that odious cap called the "Albert," with its broad black chin-strap; it had an un-English and an unsoldier-like appearance.

On the arrival of the regiment at Canterbury, from India, in 1848, it was inspected by Sir George Brown, the Adjutant-General of the Forces, when a thorough revolution was made in the clothing, and all fancy regimental specialities swept away.

The coatees had their skirts shortened, fringe on the wings reduced, and all stiffening to give them a shape prohibited, grenades removed from collars. Trousers of a slovenly cut, regulation forage caps of hideous shape and coarse quality, and "shell jackets" were ordered to be worn so loose that they were more like sacks than neatly fitting garments.

These changes were drastic ones, but had to be carried

out, and for a time the regiment did not know itself in its altered appearance. But the spirit was only scotched, not killed, for by a gradual and mysterious process, on arrival at Dublin in 1853, it had regained its former smart appearance; apparently much to the disgust of Hastings Doyle, the Adjutant-General, who in several particular items had to order a resumption of the objectionable regulation articles.

In the same year an experimental issue of Minie rifles took place, also new pattern accoutrements of the shoulder-belt only, with waist-belt and fixed "bayonet frog;" but shortly afterwards the rifles were withdrawn, to equip other regiments embarking for Turkey.

Tunics of the double-breasted pattern, with turn-back lappels, were first issued in the Crimea, and, after many alterations in shape, pattern, and quality, have assumed the tight, skimpy, choked-up appearance of those at present in use; the colour being changed from red to scarlet, the much-prized regimental button being abolished to make way for that of the "Brummagem universal pattern."

While serving in the Crimea the percussion smooth bore was exchanged for the Enfield rifle.

In Malta "fixed bayonet frogs" were exchanged for movable ones; and a new pattern cork-covered chako, with blue cloth, and furnished with a white drooping horse-hair plume, was issued.

In Dublin, in 1866, seal-skin caps were issued, and the dice band authorised for the forage cap.

In 1869 glengarries were introduced; and in 1875 the Henry-Martini was issued, together with additional ball bags and pouches to be worn on the waist-belt.

On the arrival of the first battalion from India, in 1881, the valise equipment was issued,—a decided improvement on the old knapsack.

In taking a retrospective view of the various changes and alterations, the army has much reason to congratulate itself upon serving Her Majesty Queen Victoria under the improved conditions of 1885.

Appendix No. 6.

THE COLOURS OF THE ROYAL SCOTS FUSILIERS,

PAST AND PRESENT.

I.

The Colours of the Second Battalion, Raised in 1804, and Disbanded in 1816.

THE following letter on the above subject appeared in the columns of the *Ayrshire Post* in November 1883 :—

THE ROYAL SCOTS FUSILIERS.

SIR,—As some misapprehension appears to exist regarding the colours of the Royal Scots Fusiliers, lately deposited with such impressive ceremony, and by royal hands, in St Giles' Cathedral, on Wednesday last, the following notes, partially from the 'Historical Record,' and somewhat from memory, may be found both instructive and useful to those who take a real interest in the vicissitudes of the now "county regiment" :—

The colours deposited in St Giles' are those formerly in possession of the *original* second battalion of the regiment, embodied at *Ayr* from men raised in the counties of Ayr and Renfrew on the 25th December 1804.

It remained at Ayr, after formation, until the 15th August 1806, when, marching to Portpatrick, it embarked for Belfast, where it remained for the following five years.

In September 1811 it returned to Scotland from Belfast, and was stationed at Fort-George. During the same year it furnished a strong detachment, together with volunteers from the militia, to increase the strength of the first battalion (then serving in Sicily) to a strength of 1200 rank and file.

On the 30th December 1813 the second battalion embarked from Fort-George, and landed in Holland on 18th January 1814, taking part in the attack upon Bergen-op-Zoom on the night of the 14th March. At the termination of hostilities the battalion embarked at Ostend for England, landing at Deal, and in October of the same year sailed from Gravesend to Leith, where it disembarked in the beginning of November, and proceeded to Stirling Castle.

During its stay at this station it furnished several strong reinforcements for the first battalion.

On the 13th January 1816 the battalion was disbanded at Stirling, transferring all effectives to the first battalion, then forming part of the Army of Occupation in Paris.

I have no means of ascertaining the name of the colonel commanding on its reduction, but, I presume, it must have been *Gordon*, who naturally, and as usual in those days, took possession of the colours of his late battalion, placing them in Fyvie Castle, Aberdeenshire, where they remained until so patriotically given up by his descendants for final keeping in St Giles' Cathedral, with the venerated, tattered, and glorious old standards of other Scottish regiments.

I had much pleasure in inspecting them on Wednesday. They are in a good state of preservation, and beautifully painted —not embroidered.

The colours known to be in existence, exclusive of those in use, are—

Original second battalion,	1804 to 1816—St Giles' Cathedral.
First battalion, from	1828 to 1858—Ayr Old Church.
Second battalion, from	1857 to 1864—Dumfries.

All others, as far as I am aware, have been lost sight of, viz., those carried by the regiment from 1678 to 1827.

I have heard an old story about those carried at the attack on *New Orleans*, during the War of Independence, having been preserved by a Quartermaster Sergeant Reid, who, to save them from falling into the hands of the rebels (the officers carrying them having been either killed or severely wounded), stripped them from their staves, and wrapped them around his body. Having been taken prisoner, he kept them safely until the termination of the war, and, on joining his regiment at Chatham, gave them up to the commanding-officer. For this devoted act he is said to have received a commission, and the appointment of regimental adjutant, but I do not vouch for the truth of the story. I only remember old soldiers repeating it.

Up to 1852 the colours of the Royal Scots Fusiliers bore no *honours*, but through the exertions of Lieutenant-Colonel Ainslie (who afterwards died of wounds received at Inkerman), and in justice to the regiment, "*Bladensburg*" was authorised to be emblazoned on them. Subsequently "*Alma*," "*Inkerman*," "*Sevastopol*," and

"*South Africa*," have been added; and latterly, in connection with the re-organisation of the army, it was ordered by Her Majesty, at the recommendation of a committee, that certain old victories should be added in commemoration of the distinguished services of certain regiments during the famous campaigns of the Duke of Marlborough. In accordance with this order, further additions have been made to the colours of the Royal Scots Fusiliers, which will now bear "*Blenheim*," "*Ramillies*," "*Oudenarde*," "*Malplaquet*," and "*Dettingen*," making ten in all.

Long may the old corps continue to bear these honourably, and, if dire need demands it, add to their number, in maintaining the glory of the British empire throughout the civilised world.

GEORGE GRAHAME,
Major Retired List, late Royal Scots Fusiliers.

PORTOBELLO, N.B., *17th November 1883.*

II.

Presentation of New Colours

TO THE

First Battalion of the Twenty-First Royal North British Fusiliers,

BY LADY PENNEFATHER.

[*Reprinted from the "Malta Times" of 26th January 1858.*]

AN imposing and very interesting ceremony—namely, the presentation, by Lady Pennefather, of new colours to the Twenty-first Royal North British Fusiliers—took place on the Floriana Parade, on Tuesday the 19th inst., in the presence of a large and fashionable assemblage, and a vast concourse of spectators of all classes.

The whole of the troops in garrison were on the ground, and were formed into three sides of a hollow square; the Fusiliers being in line, and in open order, and at a double distance, in the centre; the Royal Artillery, under Colonel St George, and the First Brigade, under Major-General Warren, C.B., consisting of the Fourteenth and Fifty-seventh regiments, in line of contiguous columns at right angles, on the right; the Second

Brigade in the same order, under the command of Colonel Adams, C.B., consisting of the Royal Malta Fencibles, and the Twenty-eighth regiment, on the left. Majors J. T. Dalyell and H. Gray, bearing the new colours, were posted at some distance in front of the troops.

A few minutes before the appointed hour, the Lieutenant-General, accompanied by his personal and the general staff, arrived, and was presently followed by Lady Pennefather, in a carriage, which drew up near to, but somewhat in rear of, the new colours.

Sir John now assumed the command of the division, which he called to "attention;" on which the band of the Fusiliers "trooped" in slow and quick time. On the troop being finished, the Right Rev. the Bishop of Gibraltar, attended by the Venerable Archdeacon Le Mesurier, advanced to the colours, and read impressively appropriate lessons from the Old and New Testaments, after which he consecrated the banners in a long and solemn prayer, laying his hands upon them, and blessing them.

This being done, the whole division "presented arms," the regimental bands playing the National Anthem. No. 1 Company of the Fusiliers, under Lieutenant S. H. Clerke, preceded by the regimental band, then advanced a few paces in line, and, wheeling on its left, proceeded to where the colours were posted, and then halted. The majors now advanced, and handed the colours to Lady Pennefather, who presented them to the two senior ensigns, Messrs Ker and Furlong, who received them kneeling. Here the troops cheered her ladyship enthusiastically, who acknowledged the compliment by bowing repeatedly to them.

On handing the colours to those officers, Lady Pennefather addressed them in the following brief but appropriate terms:—

"I have the honour of presenting these new colours to the Twenty-first Royal Fusiliers; and in so doing, I feel assured, whenever the regiment may be again called upon for active service, these colours will wave over it in victory. And in victory, remember mercy."

To this Lieutenant-Colonel J. R. Stuart responded thus:—

"Lady Pennefather,—Allow me, in the name of the Twenty-first Fusiliers, to thank you for the honour which you have done us in presenting us with the new colours. In 1827 those which are now furled for the last time were given to the regiment on Southsea Common by King William IV. They have been in many hands since then, and lately have had the good fortune of being carried throughout the Crimean actions. Permit me to assure your ladyship that those which we have this day received from you have been confided to trustworthy hands, and wherever it may be their fortune to serve, they will be zealously guarded.

The regiment will cherish with pride the recollection of this hour, and hereafter, when rallying round those emblems of their country's honour, we shall never forget what England, and our beloved Queen, expect from us. While a soldier of the Twenty-first Fusiliers remains, it will be no less his duty than his pride and highest ambition to keep the honour of those colours unsullied."

The officers bearing the colours now fell back on No. 1 Company, which formed the colour guard, and were received with presented arms, the band again playing "God Save the Queen." The colour guard, preceded by the regimental band playing a slow march, now advanced in line, and at open order, towards the flank of the regiment, and then wheeled to its left, disengaged its ranks, and proceeded through the open ranks of the regiment,—the colours and band in rank entire disengaging to the front, the drummers to the extreme rear, both continuing their march in the same way, the band still playing.

On the colours reaching the centre of the line of the regiment, they fell into the ranks in their proper place, which was previously occupied by the old colours, which now fell back, and were removed; but the guard and band still continued their march, until they reached the right of the line, on clearing which the guard got the word, "Front turn, halt, dress up," which brought it into its original position, and the band also resumed its proper place on the right. During this part of the ceremony the regiment, and the whole of the rest of the division, remained at presented arms, their colours being drooped, and their bands playing "God Save the Queen."

The division now came to the "shoulder," and the ranks were closed, and the Fusiliers forming "fours" closed in their centre. The brigades also closed to their respective fronts, and Sir John advanced, and in a full, clear, and distinct voice addressed the Fusiliers in the following graphical and historical speech, which we are sure will be read with interest by all who take a pride in the noble profession of arms:—

"Colonel Stuart and the Twenty-first Fusiliers,—I have been requested by Lady Pennefather to thank you for the honour you have done her in inviting her to present you your new colours. I assure you we both feel the compliment very deeply. And I venture to think she is not quite undeserving of the courtesy. She has followed my fortunes in the service for a long course of years, and in the four quarters of the world. It has been my pride before now to see my wife soothe the pillow of my sick comrades in times of pestilence; it has been my pride to see her attend the huts of my wounded comrades after battle; and I can say that she is truly the soldiers' friend.

"Soldiers, it is customary, and I think very fitting, on such an occasion as this—on such a solemn occasion—to say a few words on the previous history—the career and the manner in which it has performed its duty—of the corps. And I think I may say, without flattery, that there are few regiments in Her Majesty's army which have passed through a more honourable career, or one more marked by a constant anxiety to perform all the duties required of it than the Twenty-first Fusiliers.

"The regiment was raised in 1678, in the reign of Charles II., ten years before that glorious Revolution which established the present monarchy so firmly on the throne of England. The regiment, I say, was raised in 1678 by a great soldier of his day, the Earl of Mar, and the next year it saw its first battle. It was baptized in blood, and it distinguished itself. In 1690 the regiment served in Flanders under King William III., where it again performed its duty well. In 1702 it went to Germany with the great Duke of Marlborough. In 1704 it fought at the battle of Blenheim; its colonel, the brave Archibald Rowe, was killed at its head, forcing and cutting with his sword the palisades round the village of Blenheim. In that war also the regiment shared in the battles of Ramillies, of Oudenarde, of Malplaquet, in 1709, and performed its duty. At Malplaquet, again, it lost its colonel, the brave De Lalo, who was killed at its head. We find the regiment again in Germany. In 1743 it fought at the battle of Dettingen, under the eye of George II. Thus this corps has been engaged with the enemy under the personal command of two Kings of England, and has had the honour of fighting under the last King of England who has led his army in person to battle. In this war also the regiment was engaged at the battle of Fontenoy, in 1745, where it was distinguished, and where it suffered so severely that it became necessary to move it to Flanders. But here again duty called it into the field, and we find it in 1746 fighting at the battle of Val and at Lafeldt. After this second German war the regiment went in 1749 to Gibraltar, where it served for ten years. Gibraltar was not then, my friends, the quiet agreeable station it is at present. In that day it was open to constant attacks, was constantly threatened, and its garrison had to be constantly on the alert. In all these harassing duties, in all these watchings and alarms, the regiment was ever cheerful, ever to be depended on in the fulfilment of its duty. In 1761 the regiment was at Quebec, and was that year at the taking of Belleisle. And then, and during the succeeding year, it assisted in the military operations securing to us firmly all our Canadian possessions, and many valuable places in the West Indies and coast of America. In 1776 the regiment was engaged in the American War of Independence, and did good duty under Sir Guy Carlton; and in August of that year had hard fighting, so that in one day it lost four officers killed, and a great many non-commissioned officers and soldiers killed and wounded. The names of the officers killed on that occasion were Currie, M'Kenzie, Robertson, and Turnbull.

"In those days, my comrades, only the names of officers killed in battle were mentioned in despatches. But at present, thanks to our glorious Queen, who is ever ready to afford every indulgence, every benefit to her soldiers, to raise in every way her army, not only names of officers, but the name of every private soldier and every sergeant who distinguishes himself in action, or who dies in the performance of his duty to his Queen and country, is mentioned in the public *Gazette*, and those names go to our parishes to rejoice the hearts of our relatives and friends, or to take away half of their grief in our loss.

"In the year 1739 the regiment took its first part in the revolutionary war. That year it served at the taking of Martinique and Guadaloupe, and was particularly mentioned in despatches for its conduct at Guadaloupe by General Prescott, who commanded. In 1806 the regiment formed part of the second expedition to Egypt, under General Fraser; landed at Aboukir Bay; penetrated towards Rosetta, where there was much fighting in the desert. In 1807 the regiment was in Sicily. At that time, soldiers, there was a British army in Sicily, and the Twenty-first were actually engaged till 1810, under Sir John Stuart, fighting against Murat on the coast of Calabria and those parts, and was distinguished at the taking of Ischia and Procida, and particularly at St Stephano. In 1812 part of the corps was employed on the east coast of Spain, thus taking part in the Peninsular War, under the Duke of Wellington. Early in 1814 the regiment was at the taking of Genoa, and performed good duty, where it lost many officers and men. That same year it proceeded to America, and was at the battle of Bladensburg and the advance on Washington, doing good duty. In 1815 it was at New Orleans.

"In such a long career of military service it would be impossible not to have some reverses. There would be no honour, no renown, if all things went on smooth and successfully. We were beaten at New Orleans; but that the Twenty-first Fusiliers did their duty may be fairly gathered, I think, from this, that on that occasion they had their colonel (Colonel Reddy) killed, and left 400 officers and men killed and wounded on the field—half the regiment, but they were soon avenged, at the taking of Mobile in the Gulf of Mexico.

"In 1816 the regiment was in Paris, under the Duke of Wellington. In 1819 they went to the West Indies. In 1821 they were employed in the suppression of a dangerous insurrection in the sickly swamps of Demerara, where they performed good duty, and were particularly mentioned by the general in command,

and much promotion was given in the regiment in consequence. In 1832 the regiment went to New South Wales, and was dispersed for many years in that distant country, always cheerful, and to be depended on in its duty. In 1839 they went to Madras, and soon afterwards to Calcutta. And in 1842 marched up into the upper provinces of India, at that time, as you know, and almost ever since, the scene of stirring work with the enemy.

"But those are bygone days. Let us come to what happened yesterday—what happened under our own eyes. The Twenty-first had the honour of being engaged in the Crimean war. They landed with the army in the Crimea; they witnessed the battle of the Alma, the battle of Balaklava, the battle of Inkerman, where they again lost their colonel, the brave Ainslie; and the long and severe duty in the trenches. They were at the fall of Sebastopol, —the greatest siege in the history of the wars of the world; they were at the expedition to Kinburn; everywhere performing their duty, everywhere to be depended on.

"Men of Inkerman! does not your blood boil now when you call to mind where you fought on that glorious day, where you stood in the front of battle—in the very gap; when you now think of the heavy masses of Russians coming up against you—one after another; and how you with silence and steadiness drove them down—rolling them back in confusion and blood? On that occasion your cry was for more ammunition; never a selfish word for yourselves, though you had been at work the livelong day without food.

"Young soldiers! think of these things, determine to emulate your elders, and to gain for yourselves the same honourable distinctions they wear on their hearts. These recollections, my dear brother soldiers, do not pass us by like a summer cloud, to be forgotten; they sink deeply into the mind; and I for one do not envy that man who is not deeply moved by the remembrance of them.

"Ensigns! take these colours; they are committed to your charge in the fullest confidence. When next you are engaged march with them quietly, steadily, firmly, serenely, into the very heart of the enemy; and if you fall in the performance of your honourable office, others will at once take your place, and carry on the duty.

"Soldiers! stick to those colours. Move forward steadily, confidently; be silent, watch your officers, obey them. Obedience is discipline, and without discipline a military force is no better than an armed mob. Strike low. And I am perfectly confident, when again these colours are unfurled in war, they will, like those which have gone before, wave over your heads in victory. And when you next fight, I wish you, from the bottom of my heart, every success, and pray fervently that God Almighty may bless and prosper the corps wherever it goes."

On the conclusion of his address, Sir John was heartily cheered by the Fusiliers. The whole of the troops then marched past in review order, after which they were put through a variety of brigade evolutions by the Lieutenant-General. The division then separated into corps, and marched back to their respective barracks.

At 1.30 Sir John and Lady Pennefather, and his Excellency the Governor, and the heads of departments and their ladies, repaired to the Auberge de Castile, where they were entertained at an elegant *déjeuner* by Lieutenant-Colonel Stuart and the officers, who also, in commemoration of the occasion, gave a grand ball and supper to upwards of four hundred persons at the Auberge de Provence in the evening. Both passed off with *éclat*, and gave general satisfaction. The Twenty-first Fusiliers has always been a distinguished corps, and is deservedly popular in the island.

The following will give our readers, who may not have seen them, some idea of the colours:—The Queen's colour is represented by the "Union Jack" or national standard; the second or regimental colour is blue (the colour of the facings of the regiment), and bears on its centre the thistle (the distinguishing badge of the regiment) within a circle, having around it the well-known and very appropriate motto "*Nemo me impune lacessit.*" This is surmounted by an imperial crown. At the three corners is the Queen's cipher and crown in gold, richly embroidered; and on scrolls at each side of central badge the names of the actions in which the regiment has served.

The following is the list of guests who had the honour of receiving cards for the official *déjeuner* given by the Twenty-first on the occasion of the presentation of the new colours:—

His Excellency the Governor, the Marquis of Dalhousie, Sir John and Lady Pennefather and Staff, Vice-Admiral Sir Montagu and Lady Stopford, the Bishop, the Archdeacon, Major-General Warren and Staff, Colonel and Mrs Adams, Twenty-eighth regiment; Lieutenant-Colonel and Mrs Warre, Fifty-seventh regiment; Lieutenant-Colonel and Mrs Campbell, Seventy-first regiment; Lieutenant-Colonel Budd, Fourteenth regiment; Colonel St George, Royal Artillery; Colonel Crawley, Royal Engineers; Colonel and Mrs Greydon, Royal Artillery; Colonel and Mrs Hallewell, staff; Major Mitford, staff; Major Bayley, staff; Commissary-General and Mrs Smith; Dr Scott, P.M.O.; Captain and Mrs Mountain, Mrs St John, Mrs Pocklington, the Hon. Mrs St John, Captain and Mrs M'Donald, Mrs Stuart, Mrs Collingwood, Mrs Urquhart, Mrs Image, Mrs Grahame, and Miss Gray.

III.

Reception of the Old Colours

OF THE

First Battalion Royal Scots Fusiliers,

AT AYR, PREPARATORY TO THEIR BEING FINALLY DEPOSITED IN THE OLD CHURCH.

[*Reprinted from the "Ayr Observer."*]

A VERY interesting event happened in Ayr on Saturday evening last (30th October 1875), which, had it been publicly announced beforehand, would, we are sure, have been made the occasion of a public demonstration. As it was, a large number of people had got to know of it; and from the streets being thronged, as usual on Saturday evening, there was no lack of interested spectators.

The event we refer to was the formal reception in Ayr of the old battle-worn colours of the Twenty-first Royal North British Fusiliers, which are henceforth permanently to be retained here, as the home centre of the regiment.

According to the new army organisation scheme, instituted three years ago, the four counties of Ayr, Dumfries, Wigtown, and Kirkcudbright have been formed into what is called the Sixty-First Brigade District.

Each military district has two battalions of the line attached to it, and to these are affiliated all the militia and reserve forces within the combined counties.

The line regiments localised in this district are the first and second battalions of the Twenty-first Royal North British Fusiliers, one of which is always on foreign service, while the other is at home. Ayr has been made the depot centre of the district, and here is stationed the depot of the regiment, by which recruiting is carried on, and all the official business connected with it is transacted.

At present the second battalion of the regiment, commanded by Lieutenant-Colonel Collingwood, is stationed in the Portsdown Forts, near Portsmouth, whither all recruits at present enlisted proceed. It will remain on home station until the return of the first battalion from India, about the year 1881.

The working of the new scheme in this quarter has been attended with considerable success. Already a large number of men from the district are serving in the ranks of the Fusiliers, and, we believe, are doing well, and satisfied with their new profession. The conduct of the men stationed at the depot here has been all that could be desired, and has reflected credit upon *our own regiment.*

There is every reason to expect that, in a few years, both battalions of the Twenty-first Royal North British Fusiliers will be composed entirely of Ayrshire, Dumfriesshire, and Galloway men, and so realise the intentions of the authors of the localisation scheme. But it would help greatly to further this object if it could be arranged, as we hope it will be, to have the regiment, with its band, &c., actually stationed in the county for two or three years during its home tour of service.

As one object of the localisation scheme was to induce the people of the district to take a special interest in their own regiment, a few particulars regarding the Twenty-first Fusiliers will, we are sure, be acceptable to our readers.

The regiment was raised in 1678, by the Earl of Mar, and is, therefore, amongst the oldest of our Scottish regiments.

As mentioned by Mr D. Murray Lyon, in a series of notes on "Ayr in the Olden Time," published some time ago, the "Royal Regiment of North British Fusiliers" was garrisoned in Ayr in the year 1715, and, no doubt, helped to train the two companies of volunteers which were sent by the burgh to aid the King during the rebellion in that year. In the course of its long and distinguished history, the regiment has served in every part of the globe.

Had honourable distinctions been allowed to be worn prior to those gained in the Peninsular War, the colours of the regiment would bear, in addition to those authorised, the list of the "famous victories" gained by the Duke of Marlborough.

The original title of the regiment was the "Royal Scots Fusiliers," and although no authentic trace has been preserved of how the title came to be changed, there is reason to believe that it was at the Union with England that it received the name of the "Royal North British Fusiliers," which it still retains.

The regiment formed part of the British army that was sent to the Crimea in 1854. It distinguished itself in the battles of Alma and Inkerman, and bore the brunt of much of the hard fighting and harassing siege work during the progress of the war.

Mr Kinglake's recently published fifth volume of the history of the Crimean invasion contains a very clear account of the appearance, position, and conduct of the Fusiliers at Inkerman. Only six officers are in the regiment now who served with it in the

Crimea; but two others, at least, who were then connected with it, now occupy important positions, viz., Lieutenant-General Sir F. P. Haines, K.C.B., now commanding the Madras Army; and Major-General J. Ramsay Stuart, at present commanding the troops in Scotland. Of the survivors of the Crimean War serving at the brigade depot at Ayr, only one officer (Quartermaster Grahame) and nineteen non-commissioned officers and men remain.

The old colours, which form the principal subject of the present notice, were carried off the field of Inkerman with Colonel Ainslie, when mortally wounded. The enemy, from the conspicuous position of the colours, and knowing that they would be carried by officers and escorted by sergeants, made them a special point of aim, until they were removed from action by order of Lord Raglan. One officer, Lieutenant Hurt, was killed; another, Lieutenant Stephens, lost his arm; and a third, Lieutenant King, was very severely wounded in the neck; and some seventeen colour-sergeants and sergeants were killed and wounded while escorting them.

These colours have been in possession of the regiment since it was stationed in Ireland in 1828,—accompanying it to its various stations in England, Ireland, Scotland, Tasmania, the East Indies, the Crimea, and Malta,—but were replaced in 1858 by those now in use with the first battalion, at present serving at Rangoon, British Burmah, which were presented by Lady Pennefather, the wife of the late Sir John Pennefather, K.C.B., who commanded the forces at Malta at that time. The old colours were placed, for temporary security, in the arsenal of Edinburgh Castle, on the embarkation of the battalion for the East Indies, in February 1869.

It was felt, however, that their proper place was at the depot centre of the regiment. Accordingly, Captain J. Stevenson and an escort proceeded to Edinburgh on Saturday, and had the time-honoured colours delivered over to their custody. They arrived in Ayr with them by the 7.57 P.M. train on Saturday evening. A guard of honour, consisting of all the men of the depot of the regiment, with the drum and fife band, under the command of Captain F. M'K. Salmond, awaited their arrival at the railway station.

The colours, after having been crowned with laurel wreaths, were duly saluted, amid the vociferous cheers of a large concourse of spectators who had assembled. They were then handed by Captain F. G. Jackson and Quartermaster Grahame to Lieutenants Lambart and Meares, and were borne by them down High Street, escorted by the guard of honour, and accompanied by a dense crowd, to the barracks, where these old memorials of the many stirring events of the soldier's life amidst peace and war, will for the present find a habitation in the officers' mess-room. There they will remain until arrangements be made for a final resting-place for them, which it is hoped will be in the Old Church of Ayr, over a cenotaph or other memorial of the officers and men of the regiment who have died in the service of their country.

To commemorate this interesting event, Captains Jackson, Stevenson, and Salmond, and Quartermaster Grahame, with their wives, dined together at the mess on Saturday night.

In the course of the evening, the Crimean survivors serving at the depot were called in to drink to the old colours, under which they had served, amid scenes of hardship and danger, so many years ago.

IV.

Ceremony of Depositing the Old Colours

OF THE

First Battalion of the Royal Scots Fusiliers

IN THE OLD CHURCH AT AYR, ON THE 5TH NOVEMBER 1883.

[*Reprinted from the "Ayrshire Post."*]

On Monday forenoon a most interesting ceremony took place in the Old Church of Ayr, viz., the handing over of the old colours of the first battalion of the Twenty-first Royal Scots Fusiliers to the custody of the ministers and kirk-session of the church, with a view to their safe keeping and preservation.

The colours, which are riddled with bullet marks, and are now reduced to mere tattered shreds, were presented to the regiment in 1828, and were borne by the first battalion till the year 1858, having been carried through the Crimean campaign.

The day chosen for the ceremony of presenting the colours was a most appropriate one, being the anniversary of the battle of Inkerman. Only four of the old heroes who fought under the colours at Inkerman were present at the ceremony on Monday, viz., Sergeants Clark and Gaffney, Drum-major Maide, and Private Michael Quinn.

Those who wished to be present at the ceremony having been asked to take their seats by a quarter to eleven o'clock, at that time a large number of people had assembled in the church. Prior to the commencement of the proceedings, Mr Wilson, the organist, played several appropriate airs and marches on the organ.

A good deal of interest was manifested in the brass tablet, erected on the wall of the west transept facing the pulpit, in commemoration of the officers and men of the second battalion who fell during the recent campaign in South Africa.

At eleven o'clock the strains of "Auld Lang Syne" apprised those assembled in the church of the fact that the detachment of the regiment escorting the colours from the barracks was approaching. Colonel Allan, commanding in the Twenty-first Regimental District, was in command of the detachment.

The detachment halted at the east door of the church, and, the band having struck up "God save the Queen," the people rose up *en masse*, while the officers and men entered, bearing the colours.

The colours were then affixed to the wall of the east transept facing the pulpit, and on the opposite side of the nave from the brass tablet of the second battalion. Underneath was a small brass tablet with the following inscription :—

"The colours of the Royal Scots Fusiliers, carried by the first battalion from 1828 to 1858, through the Crimean campaign, including the battles of Alma, Balaklava, Inkerman, the siege and fall of Sebastopol, and the bombardment of Kinburn, and deposited here 1883."

The brass tablet erected by the second battalion bears the following inscription :—

"To the memory of the under-named officers and non-commissioned officers and men of the second battalion of the Twenty-first Royal Scots Fusiliers, who were killed in action, or died from wounds or disease, during the years 1879, 1881–82, in South Africa, this monument is erected, as a mark of respect and esteem, by their old comrades."

V.

Presentation of New Colours

TO THE

Second Battalion Royal Scots Fusiliers.

[*Reprinted from the "Broad Arrow" Newspaper.*]

PRESENTATION OF NEW COLOURS.

New colours were presented to the second battalion of the Twenty-first Royal Scots Fusiliers by the Duchess of Marlborough, on August 10, in the Phœnix Park, Dublin. The battalion, mustering about 550, was drawn up in line facing the Viceregal Lodge, under the command of General Sir F. W. Hamilton, K.C.B., Brevet-Colonel W. P. Collingwood, Major G. F. Gildea, Major R. W. Winsloe, Brevet-Major E. T. Bainbridge, Captain W. Thorburn, and Captain J. Whitton. The Lord-Lieutenant, the Duchess of Marlborough, Lady Georgiana and Lady Sarah Spencer Churchill, Lord Portarlington, Lord Doneraile, the Misses Baillie Cochrane (two), arrived in two open carriages, and were received with a royal salute, the colours being drooped in honour of the Viceregal party. The Duchess of Marlborough, having handed the new colours to the two senior lieutenants (Auchinlech and Dunn), said:—

"Colonel Collingwood, Officers, and Soldiers of the Royal Scots Fusiliers,—I thank you for the compliment you have paid me, in inviting me to present to you these new colours, under which you are henceforth to serve your country. I should be unworthy of the race from which I spring, and of that to which I am so closely united, if I had not a heartfelt admiration for our noble army; and I must feel a special interest in your regiment, from the fact that some of its earliest and brightest laurels were gained under the leadership of the great Duke of Marlborough.

"At the glorious battle of Blenheim it led the attack with unparalleled intrepidity. In the campaigns of William the Third, in the American War, and from that time to the Crimean campaign, with its brilliant triumphs of Alma and Inkerman, the Scots Fusiliers have gloriously distinguished themselves, and left you a noble inheritance, of which I am convinced you will not fail to prove yourselves worthy.

"Peace, the greatest of blessings, has just been preserved by our Government,—not a peace at any price, but one which rests on the heroism and efficacy of our army and navy. Never was there a time when we defended so vast a frontier. Never have we had so many nations dependent on us, but not enslaved—looking up to us for protection and good government. Never have we had greater occasion for those fleets and armies which have placed us in the foremost rank of the nations of the world. More honours than you have won under your old colours, torn as they have been by shot and shell, I cannot wish you to earn under those I am proud to deliver to you. I am confident you will sustain your reputation; and not only as soldiers, but as citizens, must you emulate the past glories of a regiment which has ever been conspicuous for high discipline, good conduct, and moral order. It is my earnest prayer that victory may attend these labours, and that success and good fortune may ever accompany you, in whatever service you may be called upon to perform."

Colonel Collingwood, advancing, replied as follows:—"May it please your grace, on behalf of the second battalion Royal Scots Fusiliers, I beg permission to express our deep gratitude for the very great mark of distinction your grace this day has been pleased to confer upon us, as no one other than the royal family could have more appropriately honoured us than the wife of a descendant of the great Marlborough. I may truly say that the feeling of loyalty and devotion to our beloved sovereign, which animates all ranks of this battalion, cannot be excelled, and they will cherish with pride their recollections of this day. The 'colours' borne for the last time to-day, and now finally furled, were presented to the battalion on its being raised, twenty years ago; and, in the name of the battalion, I feel I may faithfully promise your grace that, when and wherever these their successors are waved over the battlefield, both young and old, officers and men, will strive hard to emulate the deeds of their predecessors, and to uphold the ancient glory and reputation of the Royal Scots Fusiliers."

General Sir F. Hamilton, K.C.B., colonel of the Royal Scots Fusiliers, then said:—"I have been permitted to add a few remarks to those already made by Colonel Collingwood in the name of the battalion of which he has the honour of having command. I would, in the first place, beg to return my own personal and most sincere thanks to your grace for the great honour you have this day conferred on the corps, by the part your grace has taken in the ceremony we have just witnessed; and I am sure that every member of it must feel an additional pride in the fact that they have received their new colours from one so nearly and dearly connected with the descendant of that great soldier under whom their predecessors fought and won in the early part of last century. The several members of the corps will in course of nature pass away, but the high spirit with which it is imbued will last for ever ; so also will the remembrance of this day be handed down in the annals of the regiment to future generations, who will ever look back with pride to the fact that on two occasions—one in the time of a glorious and successful war, and the other in the time of an assured and honourable peace—they have been associated with the noble name of Marlborough. I may congratulate the Royal Scots Fusiliers that the ceremony we have just witnessed has also been honoured by the presence of his grace the Lord-Lieutenant of Ireland, and our sincere thanks are due to him for having thus honoured us. Colonel Collingwood, and officers and soldiers of the Royal Scots Fusiliers, I will conclude by the confident and sincere wish that, whenever our Queen and country may require your services against the enemy, the banners which you have had this day entrusted to your care may ever, with the help of God, wave over a victorious corps."

The colours, which are very handsome, and have, of course, the glorious records of the regiment embroidered upon them, were raised at each side of the daïs, and, the regiment having then marched past and saluted them, the ceremonial was concluded.

VI.

Depositing of the Old Colours

OF THE

Second Battalion of the Twenty-First Royal North British Fusiliers

IN THE PARISH CHURCH OF DUMFRIES.

[*Reprinted from the "Dumfries Standard" of 11th September 1878.*]

There was yesterday witnessed in Greyfriars' Church, Dumfries, a ceremony which has no precedent in the history of the burgh, —that of depositing in the parish church the old colours of a distinguished regiment, those of the second battalion of the Twenty-first Royal Scots Fusiliers.

In the recent formation of military districts in Scotland particular regiments were named for each district, which was to be considered as their headquarters. In the sub-district of which

Dumfries forms a part, the Twenty-first Royal Scots Fusiliers was the regiment which was to be connected with it.

On the recent calling out of the reserves of the Scottish Borderers Militia, the men were sent to Dublin to be associated with the Twenty-first. As they had shown great alacrity in responding to the call to be re-embodied, and had, while at Dublin, conducted themselves with propriety, and to the satisfaction of the officers of that regiment, the latter, on the occasion of new colours being presented to the Twenty-first, resolved to present their old colours to be preserved in one of the churches at Dumfries.

On Monday the colours, in charge of Captain Browne and Lieutenants Lindsel and the Hon. A. S. Hardinge, with a guard of three sergeants and a private of the Royal Scots, left Dublin by steamer for Glasgow. It was expected that the vessel would have reached that city in time for the colours to reach Dumfries yesterday forenoon, and the presentation was fixed for half-past twelve o'clock. At that hour there was a large concourse of spectators within the church. After waiting some time, the Rev. Mr Weir read a telegram to the effect that the steamer had not reached Glasgow in time for the train, and that the ceremony would not take place until half-past four, when the colours would reach Dumfries.

At a quarter to five the train arrived, and the guard in charge of the colours were met at the station by the band and staff of the Scottish Borderers, together with Captain and Adjutant Salmond and Quartermaster Irwin, two officers formerly connected with the Twenty-first; the staff presented arms, the band playing "God Save the Queen."

The party then proceeded by way of English Street and High Street to Greyfriars' Church, where Colonel Walker and other officers were in waiting at the door; also officers of the 1st D.R.V. and 5th K.R.V. The church was crowded in every part, many of the county families being present. In the magistrates' pew were Provost Smith; Bailies Smith, Wood, and Wilson; Dean Allan; Mr M'Gowan, Town Chamberlain; Mr Martin, Town Clerk; Treasurer Muirhead.

On the party reaching the church, they marched up the south passage, the men with bayonets fixed, and took their places in the cross passage in front of the pulpit, the colours being laid on the precentor's table.

The organ began to play on the party entering, and the audience rose to their feet. A hymn "For the laying up in churches of colours and standards" was then sung, and Mr Weir, who appeared in academic dress, offered up an impressive and appropriate prayer, and spoke as follows:—"About two months ago it was officially intimated to me that new colours were soon to be presented to the second battalion of the Twenty-first Royal Scots Fusiliers, and that the officers of that battalion wished to know if those who have charge of this church would approve of the old colours being placed here. I communicated this information, as I was asked to do, to the Provost and Magistrates as representing those who have civil rights in connection with this church, and to the kirk-session as representing those who have ecclesiastical rights. Both of these bodies at once expressed the great pleasure with which they heard of this proposal, and their very cordial approval of its being carried into execution.

"Now that the colours have been brought here, it is my pleasant duty to express the very great gratification which this event gives, not only to all connected with this parish church, but to very many of the inhabitants of Dumfries and its neighbourhood."

Mr Weir, after pointing out some of the reasons of this gratification, said,—

"There are circumstances that make us specially glad to receive these colours. The Twenty-first Royal Scots Fusiliers is now the regiment connected with the south-western district of Scotland, and we who live in that district have now a special right to have an interest in all that concerns it. That regiment is also one of the oldest and most distinguished in the service, and it has a history which will well add interest to everything belonging to it. It was first raised by Charles II. to assist him in opposing the Covevanters; and we may regard it as one of the picturesque incidents often wrought in the course of events, that colours belonging to a regiment which formed part ot the army of the Duke of Monmouth at Bothwell Bridge, are to-day placed in a parish church where public worship is performed after the rites of the Presbyterian Church.

"During the last century the regiment was present at much of the hard fighting which took place in every war in which the country was then engaged.

"It served under the great Duke of Marlborough; it was present in several campaigns in France; was greatly distinguished at Dettingen and Fontenoy; and it formed part of the first line of the Duke of Cumberland's army at Culloden, where it received the first shock of the charge of the Highland clans; and at the close of the century it served with great gallantry in North British America. During this century it has attained the distinctions which are recorded on its regimental colours.

"It took part in the battle of Bladensburg, in the American War, when a victory was gained over a force far superior in numbers to the Royal army; and it was present at the battles—the names of which are very familiar to this generation—of Alma, Inkerman, and Sebastopol.

"There is yet another reason why the placing of these here gives great satisfaction to the people of this district. We have been led to understand that what suggested the thought of sending them here was the alacrity with which the reserve men of the Scottish Borderers lately responded to the call of duty, and the good discipline which they showed when attached to the second battalion of the Twenty-first regiment.

"The people of the Scottish Borders are with good reason very proud of their Scottish Borderers; and they feel that any honour done to them is something at which they can all be glad. And as Dumfries is the headquarters of their regiment, and as this church is the church where the Scottish Borderers meet for Divine service, all can understand the appropriate manner in which this graceful compliment has been paid. I trust that the officers of the Royal Scots Fusiliers now present will take back to Colonel Collingwood and their brother officers, the assurance that the people of this district appreciate very highly the way in which an effort has been made to strengthen the ties between their regiment and the Scottish Borderers and the people of this district, and also the assurance that so long as these colours remain here they will be treated with every possible respect."

The seventy-second hymn was then sung, followed by the National Anthem by the choir and organ, after which Mr Weir pronounced the benediction, and the novel proceedings terminated.

The colours consist of the regimental colours proper and the "Union Jack," both dainty pieces of workmanship, but much worn and tattered, especially the latter. They will be suspended on each side of the memorial window behind the pulpit. On the colours are the names of the following battles in which the regiment has been engaged:—Bladensburg, Alma, Inkerman, and Sevastopol.

Appendix No. 7.

Bi-Centenary Reunion,

HELD AT

Ayr Barracks, on 20th September 1878,

To Commemorate the Original Formation of the Royal Scots Fusiliers in 1678.

[*Reprinted from the "Ayr Observer."*]

THE ROYAL SCOTS FUSILIERS—HISTORICAL EVENT.

On Friday last, the 20th September, the sergeants of the Royal Scots Fusiliers stationed at Ayr celebrated the two hundredth anniversary of the raising of their famous regiment by Charles, Earl of Mar.

Invitations had been forwarded to all the surviving sergeants who had served in the regiment, whose addresses were known, and the alacrity with which the various veterans acknowledged them with cordial acceptance or apologies of unfeigned regret that they were unable to attend, adds, if that were possible, an additional glory to the veneration in which the Earl of Mar's Greybreeks—for that was the old name of the regiment—is held by the present representatives.

From length, breadth, and corners of these glorious islands the heroes came. All of them had war medals, and on no better day than the 20th of September could they have met.

Most of the present generation remember Alma, and so do they,—ay, and Balaklava, Inkerman, and Sebastopol.

A magnificent banquet was spread at eight o'clock, presided over by Quartermaster-Sergeant Matier, supported by Major Hazlerigg and Captain Willoughby. The Earl of Mar having intimated that, through domestic affliction, he was debarred from attending the bi-centennial as he had intended, nevertheless an officer of the Clackmannanshire Rifle Volunteers, Mr G. Martin, was appreciated as a representative.

In addition to the sergeants past and present the wives and sweethearts were there, and since the night when at Brussels all went merry as a marriage bell no gayer scene ever gladdened the eyes of a warrior.

The Seventy-ninth Highlanders from Glasgow supplied the music, and did it well. Bandmaster M'Donald, an old Fusilier, being present as a guest, was a general favourite on the occasion. Those serving with the first battalion in India, with the second battalion in Ireland, and the old regiment throughout the remainder of the world, may remember such names as French, Geary, Harley, Talbot, Douglas, Graham, Young, Foude, Jeffrey, Flyn, Martin, Fairley, Robertson, Sweeney, Gaffney, the two Bowers, Courtney, and O'Shaughnessy, Chalmers, Sullivan, Tait, Harrington, Maley, Tyne, Preston, Russell, Benson, Sinnott, M'Cormick, Walsh, Hughes, Paterson, Hart, Derrett, Ludgate, Bailey, Clark, and Pugh, and their appearance was enough to satisfy judges that there was a good army reserve.

Each, as he entered the gay and spacious hall, was conducted to review the colours, "blood-stained, pierced, and torn," which they had so often borne on to victory; and many an undemonstrative tear was shed, and many a dearly cherished memory brought, in its entirety, forward through the years to add enjoyment to the occasion. The soldier cannot do without his comrade. The Crimea may be attached to Russia, but part of it belongs to the Royal Scots Fusiliers.

After partaking heartily of the good things of this life, the health of her Majesty, "Queen and Empress," was duly honoured. "The Services" followed in succession, the toast of the evening, however, being "The Royal Scots Fusiliers."

Dancing commenced at 10.30 P.M., and continued until Jack the piper reminded them that Johnny Cope was calling them in the morning. One and all "were happy to meet, sorry to part, and happy to meet again."

It may be remarked, in connection with this regiment, that Charles, Earl of Mar, was commissioned its first colonel on the 23d September 1678, and that the regiment was one of the first corps which obtained the distinction of being called Fusiliers. It served in the campaigns in Flanders and of Marlborough, was present at every engagement, and renewed its laurels in the Crimea.

Appendix No. 8.

PRESENTATION OF A CHALLENGE SHIELD TO THE SECOND BATTALION OF THE ROYAL SCOTS FUSILIERS.

ON the 2d of February 1875 a very interesting ceremony took place at the North Camp, Aldershot.

At noon, the 2d battalion Twenty-first Royal North British Fusiliers, under the command of Lieutenant-Colonel Collingwood, paraded for the purpose of witnessing the presentation of a challenge shield to the best shooting company for the year 1874.

Three sides of a square having been formed, the prize shield was brought forward and placed on a table. Colonel Collingwood having stepped into the square, addressed all ranks of the battalion, to the effect that he had great pleasure in assembling them on that occasion for the purpose of presenting them with a shield to be competed for annually, the winning company to retain it in their possession for one year.

When he assumed command of the battalion, he was pleased at hearing that a shield was to be offered for competition. Owing to this inducement he was glad to find that the result was an excess of average on that of the previous year, and he trusted that the average next year would be much higher. The competition was so keen, that the several companies showed very close averages. No. 1, or A Company (the victorious one), obtained 89·12 points; F Company, 88·68 points.

Mrs Collingwood then very gracefully presented the shield, and in doing so said :—"Captain Thorburn, officers, non-commissioned officers, and men of letter A Company, I have great pleasure in presenting you with this shield, which you have so ably won from so many worthy competitors. I am glad you are the first to win it, as No. 1 was the company Colonel Collingwood commanded for many years in the first battalion. I therefore feel an interest in you, and hope the shield will be honourably contested as it was last year."

Captain Thorburn, in a few appropriate and well-chosen words, thanked Colonel and Mrs Collingwood for the honour they had conferred upon his company. Next year, he said, they would again use their best endeavours to win it and to retain it.

Colonel Collingwood then called for three cheers for the victorious company, which were enthusiastically given, and the gallant company marched off proudly with their trophy, headed by the band playing the "British Grenadiers."

The shield, which was placed in the recreation room for inspection, was executed by Mr Streeter, jeweller, Bow Street, London, and is a masterpiece of design and exquisite workmanship. It is nearly two feet in diameter, and is constructed of black ebony, on which is emblazoned a number of elegantly formed devices in silver, including the honours of the regiment, "Bladensberg," "Alma," "Inkerman," and "Sebastopol." In the centre is a well-formed St Andrew's Cross, circumscribed with "Royal North British Fusiliers," and underneath, "Instituted 1874," and "Colonel Collingwood commanding." There is also neatly inscribed, "A Company, Captain Thorburn, Lieutenant Justice, and Lieutenant Alexander." And around the shield are sixteen silver plates, on which to inscribe the names of the companies which in future might succeed in becoming its possessor.

Appendix No. 9.

THE AYRSHIRE TERRITORIAL FORCE.

According to present regulations the territorial force for the counties of Ayr, Wigtown, Kirkcudbright, Dumfries, Selkirk, and Roxburgh is known as the Twenty-first Regimental District, and formed by the following corps, who are supposed, with Ayr as the centre, to be assembled for any contingency at the shortest notice :—

REGIMENTAL DISTRICT.

District Staff and Depôts of the two Line Battalions.

Line Battalions.

First and Second Battalions Royal Scots Fusiliers.

Militia Battalions.

Third Royal Scots Fusiliers, late Dumfries Militia.
Fourth Royal Scots Fusiliers, late Royal Ayr and Wigtown Militia.

Volunteer Battalions.

First Volunteer Battalion, Royal Scots Fusiliers, Roxburgh and Selkirk, including the Roxburgh Mounted Rifle Volunteer Corps.
Second Volunteer Battalion, Royal Scots Fusiliers, Kilmarnock.
Third Volunteer Battalion, Royal Scots Fusiliers, Ayr.
Fourth Volunteer Battalion, Royal Scots Fusiliers, Dumfries.
Fifth Volunteer Battalion, Royal Scots Fusiliers, Galloway, Kirkcudbright, and Wigtown.

To these may be added the "Ayrshire and Galloway Artillery Volunteers, and the regiment of Yeomen-Cavalry."

Forming, when combined, a respectable force for the protection of the homes and hearths of their native counties, and the coasts from invasion, if ever necessary.

The volunteers are a body of men of which the country is justly proud, banded together under the strongest patriotic feeling for the defence of their native soil, devoting time ungrudgingly, and striving by commendable emulation and persistent efforts to perfect themselves in their self-imposed profession.

The *Volunteer Army* is now firmly established in our midst, and gains in strength and importance year by year; and as long as such a body of men exist, danger need not be feared from a foreign foe.

The Yeomanry as a mounted force, considering the small amount of drill and instruction it is able to receive, forms a good auxiliary to the regular cavalry of the line, are well mounted, usefully armed and equipped, and present a handsome bold appearance.

The Militia battalions, being formed on the lines of regular troops, are intended as feeders for the line regiment. They do not meet this intended use as fully as they might; with few exceptions they are below their strength, and are deficient of officers, particularly those of junior rank.

A very serious defect arises from the militia regiments assembling for their annual training at different periods, which gives opportunity for a system of "bounty jumping," which means the same man being borne on the returns of several regiments, and receiving pay, clothing, rations, and bounty from each.

The recent regulations, by which the militia battalions have been affiliated to the "line county regiment," bearing its title and wearing its uniform, it is probable, will encourage *esprit de corps* between these two branches of the army, facilitate recruiting, and create a feeling of common interest in all that concerns the character, conduct, and reputation of the "County Regiments."

THE AYRSHIRE TERRITORIAL FORCE.

REGIMENTAL DISTRICT, No. 21.

The Thistle within the Circle. Saint Andrew. *The Royal Cipher and Crown. Nemo me impune lacessit.*

"Blenheim," "Ramillies," "Oudenarde," "Malplaquet," "Dettingen," "Bladensburg," "Alma," "Inkerman," "Sevastopol," "South Africa, 1879."

Line and Militia Battalions.

1st Battalion 21st Foot, Portland.
2d Battalion 21st Foot, Burmah, Madras.
3d Battalion (Scottish Borderers Militia), Dumfries.
4th Battalion (Royal Ayr and Wigtown Militia), Ayr.

Volunteer Battalions.

1st Battalion, 1st Roxburgh and Selkirk.
2d Battalion, 1st Ayrshire.
3d Battalion, 2d Ayrshire.
4th Battalion, 1st Dumfriesshire.
5th Battalion, Galloway.

Uniform—Scarlet. Facings—Blue. Agents—Messrs Cox & Co.
Head Colonel—General Sir F. W. Hamilton, K.C.B.
Lieutenant-Colonel Commanding 21st Regimental District—W. Allan.

Note.—The figures 1, 2, denote the battalion to which the officer belongs; *m*, serving with militia; *v*, with volunteers; and *s*, with names in italics, that the officer is on staff employ, and seconded, their vacancies on the effective list of the regiment being filled up; *p*, a probationer to the Indian Staff Corps.

1st and 2d BATTALIONS—LINE.

Lieut.-Colonels (4).

R. W. C. Winsloe, A.D.C., 2.
E. T. Bainbridge, 1.
F. W. Hamilton, 2.
F. G. Jackson, 1.

Majors (14).

W. A. Bridge, 1.
J. Stevenson, 2.
H. P. Law, 2.
J. Whitton, 1.
F. Mack Salmond, *s*.
F. W. Burr, 1.
E. C. Browne, 2.
H. R. C. Hewat, 2.
F. W. Douglass, *v*.
R. F. Willoughby, *v*.
J. M. Gordon, *s*.
A. J. O. Pollock, 1.
W. A. Yule, *s*.
D. Auchinlech, 2.

Captains (16).

J. H. Spurgin, 1.
F. R. H. Lambart, 2.
W. A. J. Frere, 2.
T. D. Wilson, *v*.
W. M. Duckett, *s*.
C. H. Kelly, *m*.
H. R. Alexander, *s*.
B. R. Crozier, *s*.
C. G. Knocker, *s*.
E. E. Carre.

1st and 2d Battalions—Line—*continued.*

Captains—*contd.*

A. H. Abercrombie, v.
A. W. Collings, 2.
C. Tuckey, 1.
G. A. Keef, 1.
S. F. Chichester, m.
P. W. Browne, 2.
R. C. Toogood, 1.
R. W. Blake, 1.
T. W. Fiennes, 1.
H. S. Stannel, 1.
A. W. Thorneycroft, 2.
K. E. Lean, 2.
L. N. H. D'Aeth, 1.
W. H. Bowles, 2.
J. R. M. Dalrymple Hay, 1.
H. J. Despard, adjt., 1.
F. A. L. Davidson, 2.
C. M. Eales, p.
A. B. H. Northcott, 2.
C. P. Scudamore, 2.
H. M'A. Johnston, 2.
F. V. S. Churchill, 2.
J. C. Erck, p.
D. M. Stuart, 1.
A. H. Thurburn, 2.
W. C. Walton, 2.
J. E. Vaughan, 1.
R. R. Renton, 2.
C. Bailey, 2.
H. S. Sykes, 1.
H. A. Travers, 1.
Chas. Montagu Bell, 1.
Wm. Douglas Smith, 1.
Arthur G. Baird Smith, 2.

Lieutenants (33).

W. H. Lowry, p.
C. F. Lindsell, 2.
W. A. Young, 1.
H. H. Smythe, 2.
H. J. Lermitte, 1.
Hon. A. S. Hardinge, s.
R. B. Gaisford, Adjutant, 2.
C. H. Agnew, 2.

Adjutants.

R. B. Gaisford, lieut., 2.
H. J. Despard, lieut., 1.

Quartermasters.

J. Clisham, 1.
R. J. Boddy, 2.
R. Brown, hon. captain, m.
W. J. Hancock, m.

3d BATTALION—MILITIA.

Lieut.-Colonel Commandant.

G. G. Walker, hon. col., A.D.C.

Majors (2).

J. Hatherell, hon. lt.-col.
B. T. G. Anderson.

Captains (8).

A. Hume, hon. major.
J. K. M. Witham.
R. W. Ewart.
H. Irving.
A. J. P. Johnstone.
C. V. Bayley.
J. Mackie.
J. P. K. Hannay.

Lieutenants (11).

G. Maxwell.
Sir A. D. Grierson, Bart.
J. H. M'Murdo.
W. C. S. Critchley.
W. S. Douglas.
L. B. Scott.
Oliver Rutherford.
C. H. P. Scott.
W. A. F. H. Chrichton Broune.
W. F. Carruthers.
E. J. L. Muir.

Instructor of Musketry.

Lieut. G. Maxwell.

Adjutant.

Captain C. H. Kelly (Attached from 1st Battalion).

Quartermaster.

Hon. Captain R. Browne. (Attached from 1st Battalion.)

4th BATTALION—MILITIA.

Hon. Colonel.

Sir Jas. Ferguson, Bart.

Lieut.-Colonel Commandant.

The Earl of Galloway.

Majors (2).

V. C. Sir W. J. Cuninghame, Bart., hon. col.

Captains (7).

Sir H. E. Maxwell, Bart.
W. H. Campbell.
J. F. Dalrymple Hay.
J. M. M. Morton.
C. G. Buchanan.
W. R. Dalrymple.
J. M'Haffie.

Lieutenants (13).

G. G. K. Agnew.
J. W. F. Hamilton.
G. J. Ferguson.
H. G. Wolrige Gordon.
R. L. Nugent Dunbar.
R. B. B. Christie.
W. R. Birdwood.
F. James.
T. Farquhar.
H. Sandilands.
F. C. Hunter Blair.
Sadler Hayes.
H. F. Wynn.

4th Battalion—Militia—*continued.*

Instructor of Musketry.

Captain C. G. Buchanan.

Adjutant.

Captain S. F. Chichester (Attached from 2d Line Battalion).

Quartermaster.

W. J. Hancock (Attached from 1st Line Battalion).

1st VOLUNTEER BATTALION.

The 1st Roxburgh Mounted R V.C. is attached to this Corps.

Lieutenant-Colonel.

Sir G. H. Douglas, Bart.

Major.

W. S. Elliot.

Captains.

A. L. Cochrane.
J. A. S. E. Fair.
W. Sime, hon. major.
G. Rodger.
R. Selby, hon. major.
A. E. Scougal.
J. B. Dove.
J. Carmichael.

Lieutenants.

J. Turnbull.
H. S. Murray.
J. W. Brown.
F. P. Fairbairn.
R. Innes.
A. M. Small.
D. C. Anderson.
J. Sanderson.
J. D. C. Smith.
J. Gibson.
A. Hadden.
W. Stirling.
Hon. W. G. Scott Hepburn.
C. W. Scott.
J. Richardson.

Adjutant.

Captn. J. F. M'Pherson.

Quartermaster.

J. C. Monro.

Acting-Surgeons.

J. Menzies, M.D.
J. Hume.
G. H. Turnbull.
J. S. Muir, M.B.

Acting-Chaplain.

Rev. J. C. Herdman, D.D.

2d VOLUNTEER BATTALION.

Lieutenant-Colonel.

J. Dickie.

Majors.

A. W. Faulds.
R. M. M'Kerrell.

Captains.

A. Steel.
R. Anderson.
H. M. Hight.
R. W. Patrick-Cochran.
J. R. Kelso.
W. M'Cririck.
P. Watson.
R. Fulton.
H. Crawford.

Lieutenants.

J. F. Longmuir.
A. Currie.
J. M. Stewart.
D. Walker.
P. Gorrie.
R. B. Gow.
A. Houston.
R. Lyon.

Adjutant.

Capt. A. H. Abercrombie (Attached from 1st Line Battalion).

Quartermaster.

D. Snedden.

Surgeons.

P. Munro, M.D.
J. M'Alister.

Hon. A.-Surgeon.

R. Kirkwood, M.D.

Acting-Surgeons.

W. Wilson.
A. Blair.
W. Frew, M.B.
W. Snedden, M.D.

Hon. Chaplains.

Rev. A. Hamilton.
Rev. C. Watson.

Acting-Chaplains.

Rev. M. G. Easton, D.D.
Rev. J. Grahame, B.D.

3d VOLUNTEER BATTALION.

Lieutenant-Colonel.

D. D. Whigham.

Major.

H. Ewing.

Captains.

W. Murray, hon. major.
J. G. M. Bone, hon. m.
H. Morton.
J. Patterson.
W. S. Ogilvie.
J. R. Dempster.
T. Gemmell.
W. Scott.

Lieutenants.

M. Brown.
G. Parker.
J. Chapel.
J. Morton.
D. R. Dunsmor.
D. A. White.
W. Gilmour.
J. A. Morris.
W. A. Struthers.
W. Lindsay.
J. Templeton.
J. Dodds.
J. Currie.
G. P. Walker.
J. C. Moore.
J. Andrew.

Adjutant.

Major R. F. Willoughby (Attached from 2d Line Battalion).

Quartermaster.

J. Moore.

Surgeon.

R. Dobbie, M.D., hon. surgeon-major.

Hon. A.-Surgeons.

J. Lawrence, M.D.
J. Blair.
D. Sloan.

Acting-Surgeons.

R. Girvan, M.D.
J. R. Watt, M.B.

Hon. Chaplains.

Rev. J. Thomson.
Rev. W. Corson.
Rev. J. Rankine.

Acting-Chaplains.

Rev. J. Robertson.
Rev. J. S. Robertson.

4th VOLUNTEER BATTALION.

Lieutenant-Colonel.

W. E. Malcolm.

Majors.

R. F. Dudgeon.
J. H. Dickson.

Captains.

R. Jardine.
J. Skelton.
R. Sharpe.
J. Brown.
J. B. Steuart.
J. R. MacGibbon.
P. M'Laurin.
J. Brown.

Lieutenants.

W. M'Clure.
A. Jardine.
W. Roddick.
J. S. Millar.
J. Symons.
J. R. Wilson.
J. W. Moir.
E. B. Rae.
R. Burnie.
W. J. Rae.
W. Robertson.
Aitken Malcolm.
J. Scott.
W. L. Carlyle.
J. F. C. Carruthers.

Adjutant.

Captain T. D. Wilson (Attached from 2d Line Battalion).

Quartermaster.

J. B. Riach.

Surgeon.

A. D. M'Donald, M.D.

Hon. A.-Surgeon.

W. Kay.

Acting-Surgeons.

W. J. Carlyle, M.D.
S. F. Rowan.
W. D. O'Grange, M.D.
J. Maclachlan, M.B.

Acting-Chaplains.

Rev. D. O. Ramsay.
Rev. J. Paton.
Rev. J. A. Crichton.

5th VOLUNTEER BATTALION.

Hon. Colonel.

W. K. Lawrie.

Lieutenant-Colonel.

J. G. Maitland.

Majors.

J. M. Kennedy.
W. J. Maxwell.

5th Volunteer Battalion—*continued.*

Captains.

D. Craig, hon. major.
S. Taylor, hon. major.
J. Lennox.
W. Kerr.
C. S. Phyn.
M. M'L. Harper.
J. Agnew.
J. Garrick.
J. Dunn.
J. T. Hewat.
R. Jamieson.
P. Stewart.

Lieutenants.

W. M'Lellan.
W. Craig.
J. Muir.

Adjutant.

Major F. W. Douglass (Attached from 1st Line Battalion).

Hon. A.-Surgeon.

J. Clarke.

Acting-Surgeons.

W. Johnstone.
W. Lorraine, M.D.
R. Lorraine Bell, M.B.

Hon. Chaplain.

Rev. W. M. Johnston.

Acting-Chaplains.

Rev. G. Walker, B.D.
Rev. W. Graham.
Rev. J. Mackie.

Located at Ayr, but not attached to the Territorial Regiment.

THE AYRSHIRE YEOMANRY.

Lieutenant-Colonel.

C. V. Campbell Hamilton, hon. colonel.

Major.

R. F. F. Campbell, hon. lieut.-colonel.

Captains.

W. P. Adam, hon. major.
R. M. Pollok, hon. maj.
J. Somervell, hon. major.
R. D. Murdoch.
R. Kerr.
W. D. Russell.
H. Houldsworth.
J. G. A. Baird.

Lieutenants.

W. S. Wilson.
L. G. Campbell.
J. C. C. Hamilton.
D. W. Shaw.
J. E. Dykes.
A. F. M'Adam.

Adjutant.

S. H. J. Steward (Attached from 20th Hussars).

Surgeon.

W. J. Naismith, M.D.

Veterinary Surgeon.

J. Dickie.

SCOTTISH DIVISION, R.A.—THE AYRSHIRE AND GALLOWAY ARTILLERY VOLUNTEERS.

Lieutenant-Colonel.

M. J. Stewart.

Major.

J. G. Sturrock.

Captains.

W. M'Cunn.
J. H. Turner.
T. Campbell.
A. Hamilton.
A. Guthrie.
J. Hogarth.
J. Milroy.
W. M. MacRobert.

Lieutenants.

J. Dorman.
A. M'Cubbin.
J. T. Goodwin.
J. Fleck.
A. M'Nellie.
A. M'Clymont.
J. Torrance.
R. M'Conchie.
L. Mathieson.
J. Smith.
G. O. M. Cathcart.
W. Ferguson.
J. Wight.
W. Burns.
J. M'Caig.
J. M'Harrie.

Adjutant.

Captain H. L. Murray Dunlop (Attached from Royal Artillery).

Quartermaster.

W. Ochiltree.

Surgeon.

R. B. Erskine, M.D.

Acting-Surgeons.

A. Marshall, M.D.
H. Cochrane.
J. Thomson, M.B.
R. Allan, M.B.
W. A. Caskie, M.B.
W. Moore, M.B.
A. Valentine.

Hon. Chaplains.

Rev. J. D. M'Call.
Rev. J. Sommerville.
Rev. J. Dougall.
Rev. T. Dykes.

Acting-Chaplain.

Rev. H. W. Charlton.

Appendix No. 10.

LADIES.

Mrs Gildea.

Her Majesty the Queen, by *London Gazette*, dated War Office, Horse Guards, 26th May 1884, conferred upon the above-mentioned lady the decoration of the *Royal Red Cross*, for her kindness and attention to the sick and wounded during the siege of Pretoria, the garrison of which was commanded by her husband, Lieutenant-Colonel G. F. Gildea, Second Battalion Royal Scots Fusiliers.

Mrs Hazlerigg.

This lady has established a *Soldiers' Home* at Ayr, for the use of the men of the Depôt of the Royal Scots Fusiliers, as a memorial to her late husband, Lieutenant-Colonel A. G. Hazlerigg, who took an active interest in promoting the moral and religious condition of the men under his command. This institution is known by the name of "*The Hazlerigg Soldiers' Home.*"

Mrs Grahame.

The following Testimonial, accompanied by a service of plate from the Quartermasters of the British Army, was presented to Mrs Grahame, in recognition of her husband's successful efforts in obtaining for them increased pay and rank while serving, and an improved position on retirement.

It was presented to Mrs Grahame, the rules of service not permitting her husband to accept any acknowledgment of the kind.

[COPY.]

CHATHAM, *May 1882.*

DEAR MADAM,

The Quartermasters of the British Army, with feelings of the deepest gratitude, and with a desire to express their appreciation of your husband's (Captain G. Grahame) successful efforts for the improvement of their position, beg to present you with the accompanying Testimonial, which they trust will be received as a token of their genuine and sincere admiration.

We remain,

Dear Madam,

Yours faithfully,

W. GOLDBY, *Captain, Royal Sussex Regiment, President of Committee.*
J. H. S. REID, *Major, First Life Guards.*
T. BROWN, *Quartermaster, Third Dragoon Guards.*
H. MURPHY, *Captain, Sixteenth Lancers.*
W. RICHEY, *Quartermaster, Royal Artillery.*
J. JONES, *Captain, Royal Engineers.*
J. M'BLAIN, *Captain, Scots Guards.*
R. B. JUPP, *Captain, Royal Fusiliers.*
J. R. ATKINS, *Captain, Devon Regiment.*
T. MUIR, *Captain, Bedford Regiment.*
C. PERRY, *Quartermaster, Cheshire Regiment.*
F. C. GUEST, *Quartermaster, East Lancashire Regiment.*
A. SLADE, *Captain, Prince of Wales' Volunteers.*
J. SIMPSON, V.C., *Captain, The Black Watch.*
J. REILLY, *Captain, Oxfordshire Light Infantry.*
H. HIGGINS, *Captain, Sherwood Foresters.*
W. LYNCH, *Quartermaster, Sherwood Foresters.*
G. WHITE, *Major, Royal Marine Light Infantry.*
G. T. SAVAGE, *Quartermaster, Royal Irish Rifles.*
J. E. DALTON, *Quartermaster, Commissariat and Transport Staff.*

MEMBERS OF COMMITTEE.

The plate bears the following inscription :—

Presented to

MRS G. GRAHAME,

In gratitude and admiration for the valuable services rendered

To the Quartermasters of the British Army

BY HER HUSBAND,

CAPTAIN G. GRAHAME,

Royal Scots Fusiliers.

MAY 1882.

Appendix No. II.

"FOR AULD LANG SYNE."

[*Reprinted from the Regimental Newspaper of the Second Battalion Royal Scots Fusiliers, "The Fusee," of 13th August 1884.*]

A MUCH-VALUED regimental relic arrived from home on the 5th inst., in the drum-major's staff of the old second battalion, which has been preserved in the first battalion since 1814, when the old second battalion was disbanded, and the men fit for duty transferred to the first battalion, then in Italy.

It is a very fine Malacca cane, four feet three inches in length, mounted with silver, and engraved on the top with the regimental crest—thistle and crown within the circle, with the motto "*Nemo me impune lacessit*"—and "XXI.," the number of the regiment. There is also a musical trophy of drums, bugles, and clarionets and other instruments, chased on it, and a new silver band added, bearing the following inscription:—

"This staff belonged to the old second battalion Royal Scots Fusiliers until they were disbanded in 1814, when they gave it to the first battalion, who now give it to their comrades of the second battalion as a relic of their predecessors.—*May 1884.*"

This staff was presented to the battalion on parade on the 11th inst., by the commanding-officer, with a speech appropriate to the occasion. It has been assigned a post of honour in the officers' mess-house, and will be used by the drum-major on all special parades; and we are sure that the officers, non-commissioned officers, and men of the battalion heartily thank our comrades of the first battalion for this restoration of a link in the history of our regiment, which connects us with the veterans of seventy years ago, who were of "some service to the State" in their day, and specially distinguished themselves at the desperate assault on Bergen-op-Zoom, and we feel confident that this much-prized tribute from our brother Fusiliers at home will serve to still further strengthen the feeling of good-fellowship which exists between the two battalions.

We attach a copy of the letter of thanks from Colonel Winsloe, A.D.C., to Colonel Gildea, A.D.C., which will be read with interest by all:—

"SECUNDERABAD, *5th August 1884.*

"MY DEAR GILDEA,—This day arrived the drum-major's staff of the old second battalion, and I have with my own hands placed it on the wall of the ante-room here.

"I shall show it to the men of the regiment, on parade, on the first fitting occasion, and mention the fact of its presentation in the regimental records.

"On the part of the officers, non-commissioned officers, and men of this battalion, who have asked me to do so, I write to thank yourself and all our comrades of the old battalion for the kindness which has prompted the restoring to us this relic of the old second battalion which was disbanded in 1814, and which I can assure you we will always most carefully preserve.

"Very sincerely yours,

"R. W. C. WINSLOE."

HISTORICAL RECORD

AND

REGIMENTAL MEMOIR

OF

The Royal Scots Fusiliers.

GENERAL INDEX TO CONTENTS.

www.ingramcontent.com/pod-product-compliance
Ingram Content Group UK Ltd.
Pitfield, Milton Keynes, MK11 3LW, UK
UKHW020716041026
15590UKWH00002B/697

9 781843 422402